The Geometry of Current: Unveiling the Hidden Structure of Networks

Amit

Contents

Chapter 1

Introduction

1.1 Preface

In this work we study effective resistances of finite and infinite electrical networks. In particular, we will investigate its metric properties and its connection to the underlying graph's random walk.

Consider an undirected, connected graph on a finite vertex set V with no self-loops and no multiple edges. Equipping each edge with a positive weight $c(x, y)$, such a graph becomes an electrical network if we interpret $c(x, y)$ as the conductance of a resistor between the nodes x and y. This electrical network induces an effective resistance $R(x, y)$ between every two nodes $x, y \in V$. It is a well-established result that the effective resistance is in fact a metric on V, see [23, 28, 37]. For unweighted graphs, a precise proof of this statement utilizing the connection between electric currents and random walks on graphs [10, 30] is given in [37]. An essential part of this proof is to represent the effective resistance $R(x, y)$ using the expected number of times the random walk $(X_n)_{n \in \mathbb{N}}$ on G starting in x visits x before reaching y, more precisely

$$R(x, y) = \frac{1}{c_x} \mathbf{E}_x \left[\sum_{k=0}^{\tau_y - 1} \mathbf{1}_x(X_n) \right] \tag{1.1}$$

where c_x is the sum of all edge weights attached to x.

Since every reversible, irreducible Markov chain on a finite state space V which leaves its current state with probability 1 is a random walk on some weighted graph $G = (V, c)$, see [1], this also produces a natural way to obtain a metric from such a stochastic process. Furthermore, a metric d on V admits a representation as in (1.1) if and only if it is the effective resistance of a weighted graph on V. This leads to the following questions. Which finite metric spaces (V, d) are given by effective resistances of graphs? Does there exist a concise condition in terms of the metric d? Can we extend such a condition to countably infinite metric spaces? Finally, when can a representation similar to (1.1) be extended to countably infinite metric spaces? These are the questions we seek to answer in this work. Some of the results were published in [40] and [41].

1.2 Outline

Section 1.4 establishes basic concepts, notation and standard results from graph theory, quadratic forms, function spaces and harmonic functions on graphs and probability theory.

Chapter 2 is devoted to the theory of effective resistances on finite graphs. We derive the mathematical model for our theory by considering electric currents as flows on graphs which are governed by three laws of physics: Ohm's Law, Kirchhoff's Current Law and Kirchhoff's Voltage Law. Using these laws as postulates, we show that an electric current of 1 Ampere flowing from x to y is induced by a potential function ϕ^{xy} on V which is the unique solution to the discrete Dirichlet problem

$$\Delta\phi^{xy} = \frac{1}{c_x}\mathbf{1}_x - \frac{1}{c_y}\mathbf{1}_y \tag{D}$$
$$\phi^{xy}(y) = 0 \, .$$

This leads to our definition of effective resistance $R(x, y) = \phi^{xy}(x)$. We show how effective resistances may alternatively be computed using the most general form of network reduction, the Star-mesh transform (Section 2.2) or by solving certain optimization problems regarding the graph's energy form

$$\mathcal{E}(f) = \frac{1}{2} \sum_{x,y\in V} c(x, y)(f(x) - f(y))^2,$$

see Section 2.3. More precisely, we have

$$R(x, y) = (\min\{\mathcal{E}(f) \mid f(x) = 1, f(y) = 0\})^{-1} \tag{1.2}$$

and

$$R(x, y) = \max_{\mathcal{E}(f)>0} \frac{|f(x) - f(y)|^2}{\mathcal{E}(f)}. \tag{1.3}$$

The maximum in (1.3) is attained for $f = \phi^{xy}$.

Following the approach of [37], we establish a connection between the effective resistance and the graph's random walk by proving explicit probabilistic representations (1.1) and

$$R(x, y) = \frac{1}{c_x \cdot \mathbf{P}_x[\tau_y < \tau_x^+]} \tag{1.4}$$

in Section 2.4. Building upon these, we are able to precisely compute the defects that arise in the triangle inequality of R and eventually prove that R is a metric (Section 2.5).

In Section 2.6, we show that a graph can be reconstructed from its effective resistance by solving $|V|$ linear equation systems. In particular, every graph is uniquely determined by its effective resistances.

Our main contribution to the theory of effective resistances on finite graphs are the following two characterizations of when a metric d on a finite V is the effective resistance of a graph with vertex set V. In Section 2.7, we use the aforementioned linear equation systems to produce an algebraic characterization (Theorem 2.44). For $x, y, z \in V$, let

$$A_y(x, z) := \begin{cases} 1 & , x = y = z \\ \frac{1}{2}(d(x, y) + d(y, z) - d(x, z)) & , \text{otherwise} \end{cases}$$

and $b_y(x) = 1 - \delta_y(x)$. Then, there exists a graph G with effective resistance d if and only if $\det A_y > 0$ for some $y \in V$ and the unique solution matrix $c \in \mathbb{R}^{V \times V}$ of the family of linear equation systems

$$A_y \cdot c(\cdot, y) = b_y \,, \quad y \in V$$

has only non-negative entries. In this case, $G = (V, c)$.

In Section 2.8, we develop a more geometric, necessary condition (Theorem 2.59). For any finite metric space (V, d), there exists a minimal graph $G_d = (V, c_d)$ with geodesic metric d, cf. [15, 17]. We show that if d is an effective resistance, then every (proper) cycle in G_d induces a complete subgraph. We also show that graphs of this nature can be interpreted as trees of cliques.

In Chapter 3, we review the classical theory of effective resistances on infinite graphs. Let $\operatorname{dom} \mathcal{E}$ be the space of functions with finite energy and $\operatorname{Harm} \subseteq \operatorname{dom} \mathcal{E}$ be the space of harmonic functions. Since Harm may contain non-constant functions on infinite graphs (Section 3.1), the Dirichlet problem (D) may not have a unique solution and we face some ambiguity when trying to uniquely define electric currents [13] and effective resistances. Two natural approaches lead to the notion of free and wired effective resistance R^F and R^W [19, 30], which are discussed in sections 3.2 and 3.3. Lastly, we recall both versions of Royden's decomposition [21], namely

$$\operatorname{dom} \mathcal{E} = l_0(V) + \operatorname{Harm} \tag{1.5}$$

and

$$\operatorname{dom} \mathcal{E}/\mathbb{R} = \operatorname{Fin}/\mathbb{R} \oplus \operatorname{Harm}/\mathbb{R} \tag{1.6}$$

where $\operatorname{Fin} = l_0(V) + \mathbb{R}$ and $l_0(V)$ is the closure of finitely supported functions in $\operatorname{dom} \mathcal{E}$.

Chapter 4 is concerned with the theory of resistance forms which were introduced by Kigami in [23, 24]. A resistance form $(\mathcal{F}, D[\mathcal{F}])$ on a set X is special kind of quadratic form $\mathcal{F}$ on a space $D[\mathcal{F}]$ of real-valued functions on X, see Definition 4.1. We reproduce some statements from [20, 23] and identify resistance forms on finite sets as energy forms of connected graphs (Section 4.1).

A resistance metric is a function on an arbitrary set X such that the restriction to any finite $F \subseteq X$ is the effective resistance of a graph with vertex set F. These share a one to one correspondence with resistance forms in the sense that

$$R(x, y) = \sup_{\substack{f \in D[\mathcal{F}] \\ \mathcal{F}(f) > 0}} \frac{|f(x) - f(y)|^2}{\mathcal{F}(g)} \tag{1.7}$$

for some resistance form $(\mathcal{F}, D[\mathcal{F}])$ on X. Applying Theorem 2.44, we obtain an algebraic characterization of resistance metrics (see Theorem 4.32).

In Section 4.5, we are given a resistance metric R on a countably infinite set V. This induces a sequence of growing finite graphs $G_n = (V_n, c_n)$ with invariant effective resistance $R{\restriction_{V_n}}$. We investigate the limiting behavior of the edge weights c_n. The first result of this section (Proposition 4.36) states that for fixed vertices $x, y \in V$, the sequence $c_n(x, y)$ is monotonically decreasing and thus has a limit which leads to the definition of a limit graph G_R for R.

We utilize Prohorov's theorem to show that the random walks on G_n have a weakly convergent subsequence if and only if the sequence of edge weights converges in a well-behaved manner (Theorem 4.46). This section's main result (Theorem 4.49) states that in this case, the whole sequence is weakly convergent and the weak limit is identified as the random walk of G_R. Furthermore, we show that if G_R is recurrent, R admits a probabilistic representation as in (1.1) using the limit random walk (Theorem 4.51). Since such a representation is known to exist for the free effective resistance of a recurrent graph [3], it follows that the effective resistance of G_R is exactly R.

Sections 4.6 and 4.7 are used to show that on a graph G with energy form $\mathcal{E}$, both $(\mathcal{E}, \mathrm{dom}\,\mathcal{E})$ and $(\mathcal{E}, \mathrm{Fin})$ are resistance forms whose corresponding resistance metrics are the free and wired effective resistance of G. This leads to our definition of a resistance metric of G, see Section 4.8. These are associated to resistance forms $(\mathcal{E}, D)$ such that $\mathrm{Fin} \subsetneq D \subseteq \mathrm{dom}\,\mathcal{E}$. We show that if R is a resistance metric of G, then the limit graph of R is G. The main result of this section is as follows (Corollary 4.69). If the harmonic functions in D separate the points $x \neq y$ of V, we have

$$R_D(x, y) = R^W(x, y) + \frac{1}{\mathcal{E}(h_D^{xy})} \tag{1.8}$$

where h_D^{xy} is an energy minimizer among harmonic functions in D with $h(x) = 1$ and $h(y) = 0$. Otherwise, $R_D(x, y) = R^W(x, y)$ holds. In particular, either $R^F(x, y) = R^W(x, y)$ or

$$(R^F(x, y) - R^W(x, y))^{-1} = \min \{\mathcal{E}(h) \mid h \in \mathrm{Harm}, h(x) = 1, h(y) = 0\} \tag{1.9}$$

follows.

In Chapter 5, we investigate probabilistic representations of resistance metrics of graphs. We start by reproducing lower and upper bounds for the free effective resistance

$$\frac{1}{c_x \cdot \mathbf{P}_x[\tau_y \leq \tau_x^+]} \leq R^F(x, y) \leq \frac{1}{c_x \cdot \mathbf{P}_x[\tau_y < \tau_x^+]} \tag{1.10}$$

as in [3]. On recurrent graphs (Section 5.1), we have $R^F = R^W$, i.e. there exists only one resistance metric. Moreover, both bounds in (1.10) coincide and it follows that

$$R^F(x, y) = \frac{1}{c_x \cdot \mathbf{P}_x[\tau_y \leq \tau_x^+]} = \frac{1}{c_x} \mathbf{E}_x \left[\sum_{k=0}^{\tau_y - 1} \mathbf{1}_x(X_k) \right] \tag{1.11}$$

holds.

If G is transient (Section 5.2), we produce a concise representation of the wired effective resistance in terms of the random walks Green's function $g(x, y) = \mathbf{E}_x[\sum_{k=0}^{\infty} \mathbf{1}_y(X_k)]$, namely

$$R^W(x, y) = \frac{1}{c_x}(g(x, x) - g(y, x)) - \frac{1}{c_y}(g(x, y) - g(y, y)), \tag{1.12}$$

see Corollary 5.10.

In Section 5.3, we completely characterize when the bounds in (1.10) are attained for all vertices of a graph G. Assuming that G is transient, the upper bound is attained if and only G is part of an infinite line (Theorem 5.25). The lower bound on the other hand is attained if and only if G is recurrent (Theorem 5.26). We utilize Martingale theory in Section 5.4 to produce two representations of a general resistance metric R_D of transient graphs, see Corollary 5.33. Indeed, if ϕ_D^{xy} satisfies $R_D(x,y) = \phi_D^{xy}(x) = \mathcal{E}(\phi_D^{xy})$ and $\phi_D^{xy}(y) = 0$, then $\phi_D^{xy}(X_\infty) := \lim_{n\to\infty} \phi_D^{xy}(X_n)$ exists almost surely and one has

$$R_D(x,y) = \mathbf{E}_x\left[\phi_D^{xy}(X_\infty)\right] + \frac{1}{c_x}g(x,x) - \frac{1}{c_y}g(x,y) \qquad (1.13)$$

and

$$R_D(x,y) = \frac{1}{c_x}\mathbf{E}_x\left[\sum_{k=0}^{\tau_y-1}\mathbf{1}_x(X_k)\right] + \mathbf{E}_x\left[\mathbf{1}_{\{\tau_y=\infty\}}\phi_D^{xy}(X_\infty)\right]. \qquad (1.14)$$

Since $\mathbf{P}_x[\tau_y = \infty] = 0$ on recurrent graphs, (1.14) is a direct generalization of (1.11) and holds on any graph.

1.3 Author's contribution

The following results are our main contributions to the field.

- We completely characterize algebraically when a finite metric space is the effective resistance of finite a graph (Theorem 2.44). This characterization can also be applied to resistance metrics in general (Theorem 4.32). These results were published in [40].

- We show that any finite effective resistance space has a minimal graph realization which contains no incomplete cycles (Theorem 2.59).

- For an infinite graph G, we explicitly compute the difference between a resistance metric $R_D(x,y)$ of G and its wired effective resistance $R^W(x,y)$ (Corollary 4.69). More precisely, the reciprocal of this difference is the minimal energy of a harmonic function separating the vertices x and y.

- On an infinite transient graph, we compute an explicit probabilistic representation of the potential associated with R^W in terms of the graphs green's function (Theorem 5.9). As a consequence, we get a similar representation of the graph's wired effective resistance (1.12).

- We completely characterize when the two predominant probabilistic representations for finite effective resistances hold on infinite graphs as well (Theorems 5.25 and 5.26). These results were reported in [41] and are currently being peer-reviewed.

Based on the outline, we will further clarify which parts of this book are our own work.

Sections 2.1 through 2.5 reproduce known theory of finite electrical networks found in the literature and adopt them to our context. The remainder of Chapter 2 contains our first two main contributions. We haven't seen any attempt to characterize finite effective resistance spaces in the existing mathematical literature. Sections 2.6 and 2.7 containing Theorem 2.44 were published in [40]. Section 2.8 containing Theorem 2.59 was developed in collaboration with Stefan Dück.

In Chapter 3, we state basic definitions and reproduce classic results regarding harmonic functions on infinite networks. In order to develop some intuition, we illustrate these results with our own examples.

The first half of Chapter 4, namely Sections 4.1 through 4.3 and the beginning of Section 4.4, mainly reproduce the basic theory of resistance forms as established in [20, 23, 24].

The second half of Section 4.4 containing Theorem 4.32 and some informative examples is our own contribution. Together with Section 4.5, which studies the limit graph of a resistance metric (Theorems 4.46, 4.49 and 4.51), it appeared in [40].

The main results of Sections 4.6 and 4.7, namely that the free and wired effective resistance correspond to the resistance forms $(\mathcal{E}, \mathrm{dom}\, \mathcal{E})$ and $(\mathcal{E}, \mathrm{Fin})$, respectively, are mentioned in [19, Remark 2.24]. However, we were not able to find complete proofs in the literature. The proof given in [19] for the free effective resistance seems to be incomplete, see Remark 4.57.

Our main results from Section 4.8 explicitly compute the difference in value between resistance metrics of the same graph (Theorem 4.68 and Corollary 4.69). We have not found similar notions or results in the literature.

While Section 5.1 is mainly based on existing theory found in [3], Sections 5.2 through 5.4 investigate probabilistic representations of resistance metrics on transient graphs and are our own contribution. Section 5.3 in particular is based on joint work with Stefan Bachmann which is currently being peer-reviewed [41].

We want to make this book as accessible and easy to follow as possible for readers not familiar with the theory of electrical networks. In order to achieve this, we will include proofs to almost all statements including folkloric results for which it may be difficult to find proofs in the literature. All presented proofs which are not explicitly marked otherwise are our own.

1.4   Preliminaries

In this section, we introduce basic notation and definitions as well as standard results regarding graph theory, quadratic forms, function spaces and harmonic functions on graphs and probability theory.

1.4.1   Graph Theory

Definition 1.1 (Graph)**.** A *weighted graph* G is a pair (V, c) consisting of

- a finite or countably infinite set of *vertices* $V \neq \emptyset$ and

- a *weight function* $c : V \times V \to \mathbb{R}_{\geq 0}$ such that $c(x, x) = 0$ and $c(x, y) = c(y, x)$ for all $x, y \in V$.

Furthermore, for $x \in V$ and $W \subseteq V$ let $c(x, W) := \sum_{y \in W} c(x, y)$. We say G is *locally finite* if $c_x := c(x, V) < \infty$ for all $x \in V$. We consider two vertices x, y to be adjacent if $c(x, y) > 0$ and denote the *set of edges* of G by

$$E(G) := \{(x, y) \in V \times V \mid c(x, y) > 0\}.$$

For $(x, y) \in E(G)$, define $r(x, y) := c(x, y)^{-1}$ and $r(x, y) := \infty$ otherwise. We say G is *infinite* if V is infinite. We call G *unweighted* if $c(x, y) \in \{0, 1\}$ for all $x, y \in V$.

In terms of more general Graph Theory, our graphs are undirected, allow no multiple edges and have no self-loops.

Graphs can be visualized very easily. Vertices are drawn as points or discs and edges between them are lines, see Figure 1.1 for an example. Edge weights are written next to the edges. Note that the way in which a graph is drawn is not unique.

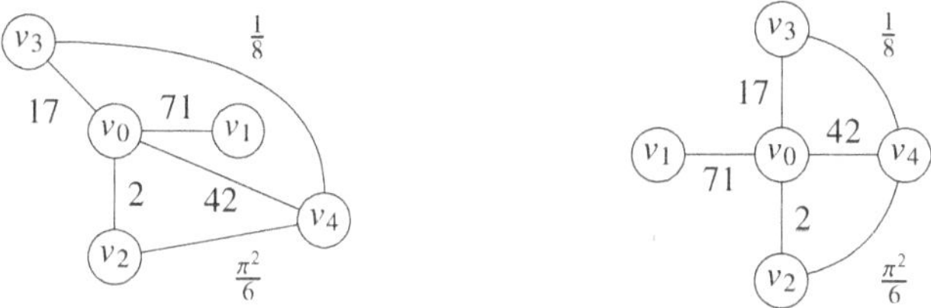

Figure 1.1: Two visualizations of the same graph.

Definition 1.2 (Finite exhaustion). Let V be a countably infinite set. A *finite exhaustion* of V is a sequence $(V_n)_{n \in \mathbb{N}}$ of subsets of V such that

1. $|V_n| < \infty$ for all $n \in \mathbb{N}$,

2. $V_n \subsetneq V_{n+1}$ for all $n \in \mathbb{N}$ and

3. $\bigcup_{n \in \mathbb{N}} V_n = V$.

If we consider infinite graphs (V, c), we will encounter sums of the form $\sum_{v \in V} f(v)$. Whenever that happens, we will implicitly assume that we have some given exhaustion $(V_n)_{n \in \mathbb{N}}$ and define

$$\sum_{v \in V} f(v) := \lim_{n \to \infty} \sum_{v \in V_n} f(v)$$

if the right-hand side exists. Note that if $f(v) \geq 0$ for all $v \in V$, then the right-hand side is independent of the choice of $(V_n)_{n \in \mathbb{N}}$ since it either converges absolutely or is infinite.

Definition 1.3 (Paths). A sequence $\gamma = (\gamma_0, \ldots, \gamma_n) \in V^{n+1}$ is called a *path (of length n)* in G if $c(\gamma_k, \gamma_{k+1}) > 0$ for all $k = 0, \ldots, n - 1$. We denote by $L(\gamma)$ the length of γ and by Γ_G the set of all paths in G. A path γ is called *simple* if it does not contain any vertex twice.

We say γ is $x \to y$ if $\gamma_0 = x$, $\gamma_{L(\gamma)} = y$ and $\gamma_k \notin \{x, y\}$ for all $k = 1, \ldots, L(\gamma) - 1$. If we need to indicate the graph G, we write $x \to_G y$. We denote by $\Gamma_G(x, y)$ the set of all paths $x \to_G y$.

For $A \subseteq V$, let

$$\Gamma_G(x, y; A) := \{\gamma \in \Gamma_G(x, y) \mid \gamma_k \in A \text{ for all } k = 0, \ldots, L(\gamma)\}$$

be the set of all paths $x \to y$ in G which only use vertices in A. A graph G is called **connected** if for any $x, y \in V$, $\Gamma_G(x, y) \neq \emptyset$.

A path γ with $n = L(\gamma) > 0$ and $\gamma_0 = \gamma_n$ is called a **cycle**. If γ satisfies $n \geq 3$ and $\gamma_i \neq \gamma_j$ for all $0 \leq i < j \leq n - 1$, we call it **proper**. A connected graph which contains no proper cycles is a **tree**.

The **weighted length (or r-length)** of γ is defined by

$$r(\gamma) := \sum_{k=0}^{L(\gamma)-1} r(\gamma_k, \gamma_{k+1}) = \sum_{k=0}^{L(\gamma)-1} \frac{1}{c(\gamma_k, \gamma_{k+1})} . \tag{1.15}$$

Remark 1.4. For the sake of brevity, we will sometimes use notations which considers paths mere sets of vertices. For example, if γ, γ' are paths, $\gamma \cap \gamma'$ is to be understood as the intersection of the sets of vertices visited by γ and γ'.

Assumptions. If not explicitly stated otherwise, we assume every occurring graph to be weighted, connected and locally finite.

Definition 1.5 (Induced subgraph). Let $G = (V, c)$ be a graph and $W \subseteq V$ a set of vertices of G. The **subgraph $G \restriction_W$ of G induced by** W is defined by $G \restriction_W := (W, c \restriction_{W \times W})$.

Note that $G \restriction_W$ is in general not connected.

1.4.2 Quadratic Forms

Definition 1.6 (Quadratic form). A mapping $q : V \to \mathbb{R}$ on a real vector space V is called a **(non-negative symmetric) quadratic form** if

1. $q(v) \geq 0$ for all $v \in V$,

2. $q(rv) = r^2 \cdot q(v)$ for all $r \in \mathbb{R}$, $v \in V$ and

3. the mapping $Q : V \times V \to \mathbb{R}$,

$$Q(v, w) = \frac{1}{2} \left(Q(u + v) - Q(u) - Q(v) \right) \tag{1.16}$$

is bilinear.

Q is called the **associated bilinear form** of q. Outside of this definition, we will abuse notation and use the same symbol for the quadratic and the bilinear form. in this notation, $q(v) = q(v, v)$ for all $v \in V$. We denote by $N(q) := \{v \in V \mid q(v) = 0\}$ the **null space** of q.

Note that, for all $v, w \in V$, we have $q(v, w) = q(w, v)$ and

$$q(v \pm w) = q(v) \pm 2q(v, w) + q(w). \tag{1.17}$$

The Cauchy-Schwarz inequality is a classic and incredibly useful tool taught and used in many undergraduate courses. Most often it is applied to infinite sums or integrals but we will include a more general form for quadratic forms for future reference. Our proof is a slight generalization of the one found in the literature for inner product spaces (see e.g. [14, Theorem 5.19]).

Theorem 1.7 (Cauchy-Schwarz inequality). *Let q be a quadratic form. Then,*

$$\forall\, v, w \in V : q(v, w)^2 \le q(v) \cdot q(w). \tag{CS}$$

Proof. Let $v, w \in V$, $\varepsilon > 0$ and $\lambda := q(v,w)/q(v)+\varepsilon$. Then,

$$0 \le q(w - \lambda v) = q(w) - 2\lambda q(v, w) + \lambda^2 q(v) = q(w) - 2\frac{q(v, w)^2}{q(v) + \varepsilon} + \frac{q(v, w)^2}{(q(v) + \varepsilon)^2} \cdot q(v)$$

$$= q(w) - q(v, w)^2 \cdot \left(\frac{1}{q(v) + \varepsilon} + \frac{\varepsilon}{(q(v) + \varepsilon)^2} \right).$$

Hence, if $q(v) \ne 0$, we have

$$q(v, w)^2 \le q(w) \cdot \left(\frac{1}{q(v) + \varepsilon} + \frac{\varepsilon}{(q(v) + \varepsilon)^2} \right)^{-1} \xrightarrow{\varepsilon \to 0} q(w) \cdot q(v).$$

If $q(v) = 0$, then

$$q(v, w)^2 \le q(w) \cdot \frac{\varepsilon}{2} \xrightarrow{\varepsilon \to 0} 0 = q(w) \cdot q(v).$$

$\square$

Corollary 1.8. *For all $v \in V$, $w \in N(q)$, we have $q(v, w) = 0$ and $q(v + w) = q(v)$. Furthermore, $N(q)$ is a linear space.*

Proof. We have

$$0 \le q(v, w)^2 \overset{\text{(CS)}}{\le} q(v) \cdot q(w) - 0.$$

Hence, $q(v, w) = 0$. This implies

$$q(v + w) = q(v) + 2q(v, w) + q(w) = q(v).$$

For $v, w \in N(q)$, $a, b \in \mathbb{R}$, we have

$$q(av + bw) = a^2 q(v) + 2ab \cdot q(v, w) + b^2 q(w) = 0.$$

Since $0 \in N(q)$, it follows that $N(q)$ is a linear space. $\square$

Example 1.9. Let $(a_n)_{n\in\mathbb{N}}$, $(b_n)_{n\in\mathbb{N}}$, $(c_n)_{n\in\mathbb{N}}$ be sequences of real numbers such that $c_n > 0$ for all $n \in \mathbb{N}$. Then,

$$\left| \sum_{n\in\mathbb{N}} a_n \cdot b_n \right| \leq \left(\sum_{n\in\mathbb{N}} c_n \cdot a_n^2 \right)^{\frac{1}{2}} \cdot \left(\sum_{n\in\mathbb{N}} \frac{1}{c_n} \cdot b_n^2 \right)^{\frac{1}{2}}. \tag{1.18}$$

For every $N \in \mathbb{N}$, the mapping $q_N : \mathbb{R}^N \times \mathbb{R}^N \to \mathbb{R}$ defined by

$$q_N(v, w) = \sum_{k=1}^{N} v_1 w_1$$

is a bilinear map and $v \mapsto q_N(v, v)$ its quadratic form. Let $a_n' = \sqrt{c_n} \cdot a_n$ and $b_n' = b_n / \sqrt{c_n}$. Furthermore, let $A_N := (a_n)_{n=1}^{N}$, $B_N = (b_n)_{n=1}^{N}$, $A_N' = (a_n')_{n=1}^{N}$ and $B_N' = (b_n')_{n=1}^{N}$. Then, we have

$$\left(\sum_{n=1}^{N} a_n \cdot b_n \right)^2 = q_N(A_N, B_N)^2 = q_N(A_N', B_N')^2 \leq q_N(A_N') \cdot q_N(B_N')$$

$$= \left(\sum_{n=1}^{N} c_n \cdot a_n^2 \right) \cdot \left(\sum_{n=1}^{N} \frac{1}{c_n} \cdot b_n^2 \right).$$

Taking the limit $N \to \infty$ implies (1.18).

For any set X, let $l(X) := \{ f : X \to \mathbb{R} \}$ be *space of real valued functions* on X. For $f, g \in l(X)$ and $a, b \in \mathbb{R}$, let

$$(af + bg)(x) = a \cdot f(x) + b \cdot g(x) \, , \; x \in X.$$

Then, $l(X)$ is a vector space. It is complete when equipped with the topology of point-wise convergence.

Proposition 1.10. *Let X be any set, $V \neq \emptyset$ a subspace of $l(X)$ which is closed w.r.t. point-wise convergence and q a quadratic form on V. Let $\|\cdot\|_q := \sqrt{q(\cdot)}$ and assume that q has the following two properties.*

1. q is lower semi-continuous, i.e. if $f_n \to f$ point-wise, then

$$q(f) \leq \liminf_{n\to\infty} q(f_n). \tag{1.19}$$

2. For every Cauchy sequence $(f_n)_{n\in N}$ w.r.t. $\|\cdot\|_q$, there exist $a_n \in N(q)$ such that $(f_n - a_n)_{n\in\mathbb{N}}$ is a Cauchy sequence w.r.t. the topology of point-wise convergence.

Then, $(V/N(q), q)$ is a Hilbert space.

Proof. Let $[v] = v + N(q)$ denote the equivalence class of v in $V/N(q)$. Since, $q(v + w) = q(v)$ for all $w \in N(q)$, $q([v]) := q(v)$ is a well-defined quadratic form on $V/N(q)$. Since $q(\cdot, \cdot)$ is bilinear and $q([v]) = 0$ implies $v \in N(q) = [0]$, it follows that q is a positive definite inner product on $V/N(q)$. In particular, $\|\cdot\|_q$ is a norm on $V/N(q)$. All that is left to show is that $\|\cdot\|_q$ induces a complete topology.

Suppose that $(f_n)_{n\in\mathbb{N}}$ is a Cauchy sequence with respect to $\|\cdot\|_q$. Then, there exist $a_n \in N(q)$ such that $(f_n - a_n)_{n\in\mathbb{N}}$ is a Cauchy sequence w.r.t. to the topology of point-wise convergence. Since $l(X)$ is complete and $V \subseteq l(X)$ is closed, there exists a function $f \in V$ such that $f_n - a_n$ converges point-wise to f. By lower semi-continuity of q, we have

$$q(f - f_n) = q(f - (f_n - a_n)) \leq \liminf_{k\to\infty} q(f_k - (f_n - a_n)) = \liminf_{k\to\infty} q(f_k - f_n)$$

for all $n \in \mathbb{N}$. Since $(f_n)_{n\in\mathbb{N}}$ is a Cauchy sequence w.r.t. $\|\cdot\|_q$, we also have $q(f_k - f_n) \to 0$ as $k, n \to \infty$. Hence,

$$\|f - f_n\|_q = \sqrt{q(f - f_n)} \to 0.$$

$\square$

For a much more general treatment of the connection between closed and lower semi-continuous forms, see [35].

1.4.3   Function Spaces

Fix a graph $G = (V, c)$. Our notation regarding the spaces of functions relating to G mostly follows the conventions of [19, 23, 30]. For the convenience of the reader, we state some basic properties and results.

Define the ***space of square summable functions on*** V as

$$l^2(V) := \left\{ f \in l(V) \mid \sum_{x\in V} c_x f(x)^2 < \infty \right\}. \tag{1.20}$$

It comes with a natural choice for an inner product

$$\langle f, g \rangle_V := \sum_{x\in V} c_x \cdot f(x) \cdot g(x) \tag{1.21}$$

making $(l^2(V), \langle \cdot, \cdot \rangle_V)$ a Hilbert space (cf. Proposition 1.10). We denote its norm by $\|\cdot\|_V$.

For $f \in l(V)$, let the ***discrete gradient of*** f be

$$(\nabla f)(x, y) := c(x, y)(f(x) - f(y)). \tag{1.22}$$

It yields a skew-symmetric function on $V \times V$, i.e. a ***flow***. Since $c(x, y) = 0$ on all $(x, y) \notin E(G)$, we can interpret ∇f as a function on $E(G)$. Conversely, any $I : E(G) \times E(G) \to \mathbb{R}$ has a natural extension to $V \times V$ by $I \restriction_{V\times V \setminus (E(G)\times E(G))} \equiv 0$. Recall that $r(x, y) = c(x, y)^{-1}$ for $(x, y) \in E(G)$ and let

$$l_-^2(E) := \left\{ I \in l(E(G) \times E(G)) \mid I(x, y) = -I(y, x), \sum_{(x,y)\in E(G)} r(x, y) \cdot I(x, y)^2 < \infty \right\}$$

$$\tag{1.23}$$

be the *space of square summable flows on* G. This space also carries a natural inner product

$$\langle I, J \rangle_E := \frac{1}{2} \sum_{(x,y) \in E(G)} r(x,y) \cdot I(x,y) \cdot J(x,y) \tag{1.24}$$

yielding the Hilbert space $(l_-^2(E), \langle \cdot, \cdot \rangle_E)$ (cf. Proposition 1.10). We denote its norm by $\|\cdot\|_E$. The factor $1/2$ is introduced because every edge $(x,y) \in E(G)$ appears twice in the above sum. We call $\mathcal{E}(I) := \langle I, I \rangle_E$ the *energy dissipation* of I.

For a flow $I \in l_-^2(E)$, we define its *divergence* $\operatorname{div} I \in l(V)$ by

$$\operatorname{div} I(x) := \frac{1}{c_x} \sum_{y \in V} I(x,y). \tag{1.25}$$

The following statements can be found in [5] without a proof which we include for completeness.

Proposition 1.11. *The operators* $\nabla : l^2(V) \to l_-^2(E)$ *and* $\operatorname{div} : l_-^2(E) \to l^2(V)$ *are bounded with operator norm less or equal to* $\sqrt{2}$ *and satisfy*

$$\langle \nabla f, I \rangle_E = \langle f, \operatorname{div} I \rangle_V. \tag{1.26}$$

Proof. For $a, b \in \mathbb{R}$, we have $(a-b)^2 \le 2(a^2 + b^2)$. Hence, for $f \in l^2(V)$, we have

$$\|\nabla f\|_E^2 = \frac{1}{2} \sum_{x,y \in V} c(x,y)(f(x) - f(y))^2 \le \sum_{x,y \in V} c(x,y)(f(x)^2 + f(y)^2)$$

$$= 2 \sum_{x \in V} c_x \cdot f(x)^2 = 2 \|f\|_V^2.$$

Let $I \in l_-^2(E)$. Using the Cauchy-Schwarz inequality, we compute

$$|(\operatorname{div} I)(x)| \le \frac{1}{c_x} \sum_{(x,y) \in E(G)} |I(x,y)| \le \frac{1}{c_x} \cdot \underbrace{\left(\sum_{(x,y) \in E(G)} c(x,y) \right)^{\frac{1}{2}}}_{=\sqrt{c_x}} \left(\sum_{(x,y) \in E(G)} \frac{1}{c(x,y)} \cdot I(x,y)^2 \right)^{\frac{1}{2}}.$$

Hence,

$$\|\operatorname{div} I\|_V^2 = \sum_{x \in V} c_x (\operatorname{div} I)(x)^2 \le \sum_{x \in V} \sum_{(x,y) \in E(G)} r(x,y) \cdot I(x,y)^2 = 2 \|I\|_E^2.$$

Furthermore, we have

$$\sum_{(x,y) \in E(G)} |f(x)| \cdot |I(x,y)| \le \left(\sum_{(x,y) \in E(G)} c(x,y) \cdot f(x)^2 \right)^{\frac{1}{2}} \left(\sum_{(x,y) \in E(G)} r(x,y) \cdot I(x,y)^2 \right)^{\frac{1}{2}}$$

$$= \sqrt{2} \cdot \|f\|_V \cdot \|I\|_E < \infty$$

by another application of Cauchy-Schwarz. Hence, changing the order of summation is justified in the following computation.

$$\langle \nabla f, I \rangle_E = \frac{1}{2} \sum_{(x,y)\in E(G)} r(x,y) \cdot c(x,y) \cdot (f(x) - f(y)) \cdot I(x,y)$$

$$= \frac{1}{2} \sum_{(x,y)\in E(G)} f(x) \cdot I(x,y) + \frac{1}{2} \sum_{(x,y)\in E(G)} f(y) \cdot I(y,x))$$

$$= \sum_{(x,y)\in E(G)} f(x) \cdot I(x,y) = \sum_{x\in V} \left(c_x \cdot f(x) \cdot \frac{1}{c_x} \sum_{(x,y)\in E(G)} I(x,y) \right) = \langle f, \operatorname{div} I \rangle_V$$

$\square$

We define the **Graph Laplacian** $\Delta : l^2(V) \to l^2(V)$ by $\Delta = \operatorname{div} \circ \nabla$. Explicitly, this yields

$$(\Delta f)(x) = f(x) - \sum_{y\in V} \frac{c(x,y)}{c_x} f(y) = \frac{1}{c_x} \sum_{y\in V} c(x,y)(f(x) - f(y)) . \tag{1.27}$$

$(\Delta f)(x)$ measures how much the value $f(x)$ differs from the weighted average of values of f computed over all adjacent vertices. By Proposition 1.11, Δ is self-adjoint on $l^2(V)$. We say a function f is **harmonic** if $\Delta f \equiv 0$.

Proposition 1.12. *If G is connected, then all harmonic functions in $l^2(V)$ are constant.*

Proof. Let $f \in l^2(V)$. Then,

$$\|\nabla f\|_E = \langle \nabla f, \nabla f \rangle_E = \langle f, \Delta f \rangle_V = 0$$

implies that $\nabla f \equiv 0$. For $(x, y) \in E(G)$, we have

$$0 = (\nabla f)(x, y) = c(x,y)(f(x) - f(y))$$

if and only if $f(x) = f(y)$. Hence, f has constant value along every path $\gamma \in \Gamma_G$. Since G is connected, there exists a path $x \to y$ for every $x, y \in V$. $\square$

Energy Spaces

The **energy form** $\mathcal{E}$ of a graph $G = (V, c)$ is a quadratic form $l(V) \to \mathbb{R} \cup \{\infty\}$ defined by

$$\mathcal{E}(f) := \begin{cases} \langle \nabla f, \nabla f \rangle_E & , \nabla f \in l^2(E) \\ \infty & , \nabla f \notin l^2_-(E) \end{cases} . \tag{1.28}$$

In this sense, we have

$$\mathcal{E}(f) = \frac{1}{2} \sum_{x,y\in V} c(x,y)(f(x) - f(y))^2 \tag{1.29}$$

for all $f \in l(V)$. We write $\mathcal{E}_G$ if we need to indicate the graph G which the energy form belongs to.

On the *space of functions of finite energy* $\operatorname{dom}\mathcal{E} := \{f \in l(V) \mid \mathcal{E}(f) < \infty\}$, its associated bilinear form $\mathcal{E}(f,g) := \langle \nabla f, \nabla g \rangle_E$ is well-defined and satisfies

$$\mathcal{E}(f,g) = \frac{1}{2} \sum_{x,y \in V} c(x,y) \cdot (f(x) - f(y)) \cdot (g(x) - g(y)). \tag{1.30}$$

By Proposition 1.11, we have $l^2(V) \subseteq \operatorname{dom}\mathcal{E}$. Since $\operatorname{dom}\mathcal{E} = \{f \in l(V) \mid \nabla f \in l_-^2(E)\}$, we can extend the domain of Δ to be $\operatorname{dom}\mathcal{E}$. Note however, that $\Delta : \operatorname{dom}\mathcal{E} \to l^2(V)$ is not symmetric in general.

Define $m : \mathbb{R} \to [0,1]$ by

$$m(x) = \min(\max(x,0),1), \; x \in \mathbb{R}. \tag{1.31}$$

Then, $|m(x) - m(y)| \leq |x - y|$ for all $x, y \in \mathbb{R}$. We note some basic properties of $\mathcal{E}$ in the following lemma and leave the proof as an exercise.

Lemma 1.13. *For $f, g \in l(V)$, we have*

1. $\mathcal{E}(f) \geq 0$,

2. $\mathcal{E}(f) = 0$ *if and only if f is constant,*

3. $\mathcal{E}(f + k) = \mathcal{E}(f)$ *for any $k \in \mathbb{R}$ and*

4. $\mathcal{E}(m(f)) \leq \mathcal{E}(f)$.

Proposition 1.14. *For $f \in l^2(V)$, $g \in \operatorname{dom}\mathcal{E}$, we have $\mathcal{E}(f,g) = (f, \Delta g)_V$.*

Proof. Let $f \in l^2(V)$, $g \in \operatorname{dom}\mathcal{E}$. Then,

$$\sum_{x,y \in V} c(x,y) \cdot |f(x)| \cdot |g(x) - g(y)| \leq \|f\|_V \cdot \|\nabla g\|_E < \infty$$

by Cauchy-Schwarz. Hence, we have absolute convergence and can freely change the order of summation of

$$\sum_{x,y \in V} c(x,y) \cdot f(x) \cdot (g(x) - g(y))$$

in order to compute

$$\begin{aligned}
\mathcal{E}(f,g) &= \frac{1}{2} \sum_{x,y \in V} c(x,y)(f(x) - f(y))(g(x) - g(y)) \\
&= \frac{1}{2} \sum_{x,y \in V} c(x,y)f(x)(g(x) - g(y)) - \frac{1}{2} \sum_{x,y \in V} c(x,y)f(y)(g(x) - g(y)) \\
&= \frac{1}{2} \sum_{x,y \in V} c(x,y)f(x)(g(x) - g(y)) + \frac{1}{2} \sum_{x,y \in V} c(y,x)f(y)(g(y) - g(x)) \\
&= \sum_{x,y \in V} c(x,y)f(x)(g(x) - g(y)) \\
&= \sum_{x \in V} f(x) \cdot \left(\sum_{y \in V} c(x,y)(g(x) - g(y)) \right) \\
&= \sum_{x \in V} f(x) \cdot c_x \cdot (\Delta g)(x) = (f, \Delta g)_V.
\end{aligned}$$

$\square$

For $A \subseteq V$, let $\mathbf{1}_A : V \to \{0, 1\}$ be the ***characteristic function of*** A, i.e.

$$\mathbf{1}_A(x) = \begin{cases} 1 & , x \in A \\ 0 & , x \notin A \end{cases} \tag{1.32}$$

and define $\mathbf{1}_x := \mathbf{1}_{\{x\}}$ for $x \in V$. We write $\mathbf{1}$ and $\mathbf{0}$ for the constant functions of value 1 and 0, respectively. Applying Proposition 1.14 to $f = \mathbf{1}_x \in l^2(V)$ yields

$$\mathcal{E}(\mathbf{1}_x, g) = c_x \cdot (\Delta g)(x). \tag{1.33}$$

In particular, we have

$$\mathcal{E}(\mathbf{1}_x, \mathbf{1}_x) = c_x \tag{1.34}$$

and, for $x \neq y$,

$$\mathcal{E}(\mathbf{1}_x, \mathbf{1}_y) = -c(x, y). \tag{1.35}$$

Together with $c(x, x) = 0$ for all $x \in V$, this implies following proposition.

Proposition 1.15. *Let $G_1 = (V, c_1)$ and $G_2 = (V, c_2)$ be two graphs with the same vertex set and energy forms $\mathcal{E}_1$ and $\mathcal{E}_2$, respectively. Then, $\mathcal{E}_1 = \mathcal{E}_2$ if and only if $G_1 = G_2$.*

Proposition 1.16. *For $x \in V$, $f \in l(V)$, let*

$$(\mathcal{E} + \delta_x^2)(f) := \mathcal{E}(f) + f(x)^2. \tag{1.36}$$

Then, $\mathrm{H}_x := (\mathrm{dom}\, \mathcal{E}, \mathcal{E} + \delta_x^2)$ and $\mathcal{H}_{\mathcal{E}} := (\mathrm{dom}\, \mathcal{E}/\mathbb{R}, \mathcal{E})$ are Hilbert spaces.

We leave the proof as an exercise. We recommend using Proposition 1.10 together with the following proposition and the semi-continuity of $\mathcal{E}$ which we will show at a later point (cf. Lemma 4.54).

Proposition 1.17. *If $f_n \to f$ in H_x for some $x \in V$, then $f_n \to f$ point-wise.*

Proof. We have $\mathcal{E}(f_n - f) \to 0$ and $f_n(x) \to f(x)$. Let $y \in V$ and $(x_0, \ldots, x_m)$ be a path $x \to y$ in G. Then, $c(x_k, x_{k+1}) > 0$ and

$$c(x_k, x_{k+1})(f_n(x_{k+1}) - f_n(x_k) - (f(x_{k+1}) - f(x_k)))^2 \leq \mathcal{E}(f_n - f) \to 0$$

for all $k = 0, \ldots, m - 1$. Hence, $f_n(x_{k+1}) - f_n(x_k) \to f(x_{k+1}) - f(x_k)$ as $n \to \infty$. It follows that

$$f_n(y) = f_n(x) + \sum_{k=0}^{n-1} (f_n(x_{k+1}) - f_n(x_k)) \to f(x) + \sum_{k=0}^{n} (f(x_{k+1}) - f(x_k)) = f(y).$$

$\square$

The above statement implies that a sequence $(f_n)_{n \in \mathbb{N}} \subseteq \mathrm{dom}\, \mathcal{E}$ converges in H_x if and only if it converges H_y. Hence, the norms of H_x and H_y are strongly equivalent.

Lemma 1.18. *The bounded functions of* dom $\mathcal{E}$ *are dense in* H_x. *More precisely, for every* $f \in$ dom $\mathcal{E}$, *there exist bounded* $f_n \in$ dom $\mathcal{E}$ *such that* $f_n \to f$ *in* H_x.

Proof. For $N \in \mathbb{N}$, let $m_N : \mathbb{R} \to [-N, N]$ be the cut-off at $-N$ and N, i.e.

$$m_N(x) := \min(\max(-N, x), N) \, , \, x \in \mathbb{R}.$$

Then, m_N is a contraction, i.e. $|m_N(x) - m_N(y)| \leq |x - y|$ for all $x, y \in \mathbb{R}$ and it follows that $\mathcal{E}(m_N(f)) \leq \mathcal{E}(f)$ for all $f \in l(V)$. Let $f \in$ dom $\mathcal{E}$ and $f_n := m_n(f)$. Then, $f_n \in$ dom $\mathcal{E}$, it is bounded and $f_n \to f$ point-wise. By Lebesgue's Dominated Convergence Theorem [34, Theorem 11.2], we have $\mathcal{E}(f_n - f) \to 0$ and $\|f_n - f\|_{H_x} \to 0$ follows. $\square$

The following is a general theorem from Hilbert space theory and can be found in [3, Theorem 2.16].

Theorem 1.19. *Let* $A \neq \emptyset$ *be a closed, convex subset of a Hilbert space* H. *Then,*

1. *There exists a unique element* $m \in A$ *with minimal norm.*

2. *If* $(f_n)_{n \in \mathbb{N}} \subset A$ *is a sequence with* $\|f_n\|_H \to \|m\|_H$, *then* $f_n \to m$ *in* H.

3. *For any* $h \in H$ *and* $\varepsilon > 0$ *such that* $m + d \cdot h \in A$ *for all* $|d| < \varepsilon$, *we have* $\langle m, h \rangle = 0$.

1.4.4 Probability Theory

We will rely heavily on the theory of (time-homogeneous, irreducible and reversible) Markov chains. If not already the case, we recommend [1, 11, 32] to become more familiar with the subject.

Random Walk on G

Definition 1.20 (Markov chain). Let S be a countable set and $p : S \times S \to [0, 1]$ such that $\sum_{t \in S} p(s, t) = 1$ for all $s \in S$. A sequence of random variables $(X_n)_{n \in \mathbb{N}}$ with values in S is called a **Markov chain** on S with **transition probabilities** p and **distributions** $(\mathbf{P}_s)_{s \in S}$ if the following conditions are met.

1. For each $s \in S$, $\mathbf{P}_s$ is a probability distribution on the **sequence space** $\Omega = S^{\mathbb{N}}$ with

$$\mathbf{P}_s[X_0 = s] = 1. \tag{1.37}$$

2. For all $n \geq 1$, $s_0, s_1, \ldots, s_n \in S$, we have

$$\mathbf{P}_{s_0}[X_1 = s_1, \ldots, X_n = s_n] = \prod_{k=0}^{n-1} p(s_k, s_{k+1}). \tag{1.38}$$

We denote by $\mathbf{E}_x$ the expectation of $\mathbf{P}_x$, i.e. $\mathbf{E}_x[f] = \int_\Omega f \, d\mathbf{P}_x$.

Remark 1.21. If not explicitly stated otherwise, we consider S to be endowed with the discrete metric $d_{\mathrm{disc}}(s,t) = 1 - \mathbf{1}_s(t)$. The induced Borel-$\sigma$-field is the power set 2^S of S. The sequence space Ω then carries the corresponding product topology and its Borel-σ-field. The product topology is metrizable and is induced by the following metric (see [12, Theorem 4.2.2])

$$\rho(\omega, \omega') = \sum_{k=0}^{\infty} 2^{-k} \cdot d_{\mathrm{disc}}(\omega_k, \omega'_k). \tag{1.39}$$

Markov chains have the ***Markov property*** which can be expressed in many ways. We will mostly use two basic versions. If $(X_n)_{n \in \mathbb{N}}$ is a Markov chain on S, $A_1, \ldots, A_{k-1} \subseteq S$ and $s, s_k, s_{k+1} \in S$, then

$$\mathbf{P}_s[X_{k+1} = s_{k+1} \mid X_k = s_k, X_{k-1} \in A_{k-1}, \ldots, X_1 \in A_1] = \mathbf{P}_s[X_{k+1} = s_{k+1} \mid X_k = s_k].$$
$$\tag{MP}$$

Furthermore,

$$\mathbf{P}_s[X_{k+1} = s_{k+1} \mid X_k = s_k] = p(s_k, s_{k+1}).$$

Both equalities follow directly from (1.38). For the second version, let $\Phi : S^{\mathbb{N}_0} \to \mathbb{R}$ be measurable, $s \in S$ and $n \in \mathbb{N}_0$. Then,

$$\mathbf{E}_s[\Phi(X_n, X_{n+1}, \ldots)] = \mathbf{E}_s[\mathbf{E}_{X_n}[\Phi(X_0, X_1, \ldots)]] \tag{MP2}$$
$$= \sum_{t \in S} \mathbf{P}_s[X_n = t] \cdot \mathbf{E}_t[\Phi(X_0, X_1, \ldots)]$$

holds (cf. [11, Theorem 6.3.1]).

One can show that for given transition probabilities p, there always exists a Markov chain as above. This is mainly due to Kolmogorov's extension theorem, see [11, Chapter 6].

Definition 1.22 (Random walk on G). The ***random walk on*** G is the Markov chain $(X_n)_{n \in \mathbb{N}}$ on V with transition probabilities

$$p(y, z) = \frac{c(y, z)}{c_y} , \, y, z \in V \tag{1.40}$$

and distributions $(\mathbf{P}_x)_{x \in V}$. For $A \subseteq V$, let

$$\iota_A - \inf \{ k \geq 0 \mid X_k \in A \} \tag{1.41}$$

be the ***first time the random walk visits*** A. Furthermore, define

$$\tau_A^+ := \inf \{ k \geq 1 \mid X_k \in A \} . \tag{1.42}$$

For $v \in V$, we will use the short-hand notation $\tau_v := \tau_{\{v\}}$ and $\tau_v^+ := \tau_{\{v\}}^+$.

A vertex $x \in V$ is called ***recurrent*** if $\mathbf{P}_x[\tau_x^+ < \infty] = 1$ and ***transient*** otherwise. Since we consider connected graphs, the random walk $(X_n)_{n \in \mathbb{N}}$ is an irreducible Markov chain. Hence, either every vertex is recurrent or every vertex is transient. If all vertices are recurrent, we say G is ***recurrent***.

Green's Function

For $n \in \mathbb{N}_0$, let

$$V_A^n := \sum_{k=0}^{n} \mathbf{1}_A(X_k) \tag{1.43}$$

be the ***number of visits of the random walk to vertices in*** A ***during the first*** n ***steps.*** We will use the short-hand notation $V_A := V_A^\infty$ and, for $v \in V$, $V_v^n := V_{\{v\}}^n$.

The ***Green's function*** of $(X_n)_{n \in \mathbb{N}}$ is defined by

$$g(x, y) := \mathbf{E}_x[V_y]. \tag{1.44}$$

Lemma 1.23. *Let* $x \in V$ *and* $k \in \mathbb{N}^+$. *Then,* $\mathbf{P}_x[V_x \geq 0] = 1$ *and*

$$\mathbf{P}_x[V_x \geq k] = \mathbf{P}_x[\tau_x^+ < \infty]^{k-1}. \tag{1.45}$$

If $y \in V$, $x \neq y$, *we also have* $\mathbf{P}_x[V_y \geq 0] = 1$ *and*

$$\mathbf{P}_x[V_y \geq k] = \mathbf{P}_x[\tau_y < \infty] \cdot \mathbf{P}_y[\tau_y^+ < \infty]^{k-1}. \tag{1.46}$$

Proof. Let $x \in V$. We have $\mathbf{P}_x[V_x \geq 0] = 1$ by definition of V_x and we will use induction over k to prove (1.45). Since $X_0 = x$ holds $\mathbf{P}_x$-almost surely, $\mathbf{P}_x[V_x \geq 1] = 1 = \mathbf{P}_x[\tau_x^+ < \infty]^0$ follows. For $k > 1$, we use (MP) to compute

$$\mathbf{P}_x[V_x \geq k] = \sum_{n=1}^{\infty} \mathbf{P}_x[V_x \geq k \mid \tau_x^+ = n] \cdot \mathbf{P}_x[\tau_x^+ = n]$$

$$= \sum_{n=1}^{\infty} \mathbf{P}_x[V_x \geq k \mid X_n = x, X_l \neq x \ \forall \ 0 < l < n] \cdot \mathbf{P}_x[\tau_x^+ = n]$$

$$= \sum_{n=1}^{\infty} \mathbf{P}_x[V_x \geq k - 1] \cdot \mathbf{P}_x[\tau_x^+ = n] = \mathbf{P}_x[V_x \geq k - 1] \cdot \sum_{n=1}^{\infty} \mathbf{P}_x[\tau_x^+ = n]$$

$$= \mathbf{P}_x[V_x \geq k - 1] \cdot \mathbf{P}_x[\tau_x^+ < \infty] = \mathbf{P}_x[\tau_x^+ < \infty]^{k-1}.$$

where the last equality follows by induction.

For $y \in V$, $y \neq x$, $\mathbf{P}_x[V_y \geq 0] = 1$ holds by definition of V_y. For $k \geq 1$, we compute analogously to before

$$\mathbf{P}_x[V_y \geq k] = \sum_{n=0}^{\infty} \mathbf{P}_x[V_y \geq k \mid \tau_y = n] \cdot \mathbf{P}_x[\tau_y = n]$$

$$= \mathbf{P}_y[V_y \geq k] \cdot \sum_{n=0}^{\infty} \mathbf{P}_x[\tau_y = n] = \mathbf{P}_y[V_y \geq k] \cdot \mathbf{P}_x[\tau_y < \infty]$$

and (1.46) follows from (1.45).   □

Lemma 1.24. *Assume that* G *is transient. For* $x, y \in V$, $x \neq y$, *we have*

$$g(x, x) = \frac{1}{1 - \mathbf{P}_x[\tau_x^+ < \infty]} \tag{1.47}$$

and

$$g(x, y) = \frac{\mathbf{P}_x[\tau_y < \infty]}{1 - \mathbf{P}_y[\tau_y^+ < \infty]} \, . \tag{1.48}$$

In particular, $g(x, y) < \infty$ if and only if G is transient.

Proof. By Lemma 1.23, we have

$$\mathbf{P}_x[V_x = k] = \mathbf{P}_x[V_x \geq k] - \mathbf{P}_x[V_x \geq k + 1] = \mathbf{P}_x[\tau_x^+ < \infty]^{k-1} \cdot (1 - \mathbf{P}_x[\tau_x^+ < \infty]).$$

for $k \geq 1$. For convenience, let $p := \mathbf{P}_x[\tau_x^+ < \infty]$. Since G is transient, $V_x < \infty$ holds $\mathbf{P}_x$-almost surely and $p < 1$. Hence,

$$g(x, x) = \mathbf{E}_x[V_x] = \sum_{k=0}^{\infty} k \cdot \mathbf{P}_x[V_x = k] = (1 - p) \cdot \sum_{k=1}^{\infty} k \cdot p^{k-1} = (1 - p) \cdot \frac{\mathrm{d}}{\mathrm{d}p} \left(\sum_{k=0}^{\infty} p^k \right)$$

$$= (1 - p) \cdot \frac{\mathrm{d}}{\mathrm{d}p} \left(\frac{1}{1 - p} \right) = (1 - p) \cdot \frac{1}{(1 - p)^2} = \frac{1}{1 - \mathbf{P}_x[\tau_x^+ < \infty]} \, .$$

Note that if G were recurrent, we have $p = 1$ and $g(x, x) = \mathbf{E}_x[V_x] = \infty$ follows.

For $y \in V$, $x \neq y$, we analogously have

$$\mathbf{P}_x[V_y = k] = \mathbf{P}_x[\tau_y < \infty] \cdot \mathbf{P}_y[\tau_y^+ < \infty]^{k-1} \cdot (1 - \mathbf{P}_y[\tau_y^+ < \infty])$$

and

$$g(x, y) = \mathbf{E}_x[V_y] = \mathbf{P}_x[\tau_y < \infty] \cdot \sum_{k=0}^{\infty} \mathbf{P}_y[\tau_y^+ < \infty]^{k-1} = \frac{\mathbf{P}_x[\tau_y < \infty]}{1 - \mathbf{P}_y[\tau_y^+ < \infty]}$$

follows as above. Again, recurrence of G would imply $g(x, y) = \infty$. $\qquad\square$

Using the same methods as in the proof above, one can show the following statement about the second moment of V_y.

Lemma 1.25. *Assume that G is transient. For $x, y \in V$, $x \neq y$, we have*

$$\mathbf{E}_x[V_x^2] = \frac{1 - \mathbf{P}_x[\tau_x^+ < \infty]^2}{(1 - \mathbf{P}_x[\tau_x^+ < \infty])^3} \tag{1.49}$$

and

$$\mathbf{E}_x[V_y^2] = \mathbf{P}_x[\tau_y < \infty] \cdot \frac{1 - \mathbf{P}_y[\tau_y^+ < \infty]^2}{(1 - \mathbf{P}_y[\tau_y^+ < \infty])^3} \, . \tag{1.50}$$

For any $f \in l(V)$, we have

$$(\Delta f)(x) = f(x) - \mathbf{E}_x[f(X_1)]. \tag{1.51}$$

For $f = g(\cdot, x)$, the Markov property (MP2) implies

$$g(y, x) = \mathbf{E}_y\left[\sum_{k=0}^{\infty} \mathbf{1}_x(X_k) \right] = \mathbf{1}_x(y) + \mathbf{E}_y\left[\sum_{k=1}^{\infty} \mathbf{1}_x(X_k) \right]$$

$$= \mathbf{1}_x(y) + \sum_{z \in V} p(y, z) \cdot \mathbf{E}_z\left[\sum_{k=0}^{\infty} \mathbf{1}_x(X_k) \right] = \mathbf{1}_x(y) + \mathbf{E}_y[g(X_1, x)]$$

and it follows that

$$\Delta g(\cdot, x) = \mathbf{1}_x. \tag{1.52}$$

Measure Theory

We recall some notions and theorems from measure theory. Proofs can be found in [34, Chapter 16]. Throughout, let $(\Omega, \mathcal{A}, \mu)$ be a σ-finite measure space.

Definition 1.26 (Convergence in measure). We say a sequence $(f_n)_{n \in \mathbb{N}}$ of measurable functions converges to a measurable function f **_locally in measure_** if

$$\forall\, \varepsilon > 0 \,\forall\, A \in \mathcal{A}, \mu(A) < \infty : \lim_{n \to \infty} \mu\left(\{|f_n - f| > \varepsilon\} \cap A\right) = 0. \tag{1.53}$$

Note that μ-almost everywhere convergence implies convergence locally in measure.

Definition 1.27 (Uniformly integrable). A sequence $(f_n)_{n \in \mathbb{N}}$ of measurable functions is called **_uniformly integrable_** if

$$\forall\, \varepsilon > 0 \,\exists\, w_\varepsilon \in \mathcal{L}_+^1(\mu) : \sup_{n \in \mathbb{N}} \int_{\{|f_n| > w_\varepsilon\}} |f_n|\, \mathrm{d}\mu < \varepsilon. \tag{1.54}$$

If $\mu(\Omega) < \infty$, this is equivalent to

$$\lim_{M \to \infty} \sup_{n \in \mathbb{N}} \int_{\{|f_n| > M\}} |f_n|\, \mathrm{d}\mu = 0 \tag{1.55}$$

and to the existence of an increasing, convex function $H : [0, \infty) \to [0, \infty)$ such that $\lim_{t \to \infty} H(t)/t = \infty$ and

$$\sup_{n \in \mathbb{N}} \int H(|f_n|)\, \mathrm{d}\mu < \infty. \tag{1.56}$$

Suppose that $(\Omega, \mathcal{A}, \mu)$ is a probability space. Then, a sequence $(f_n)_{n \in \mathbb{N}}$ is uniformly integrable if there exists $p > 1$ such that

$$\sup_{n \in \mathbb{N}} \int |f_n|^p\, \mathrm{d}\mu < \infty$$

since $t \mapsto t^p$ is increasing, convex and satisfies $t^p/t = t^{p-1} \to \infty$.

Theorem 1.28 (Vitali's convergence theorem). *Let $(\Omega, \mathcal{A}, \mu)$ be a σ-finite measure space and $p \in [1, \infty)$. Furthermore, let $(f_n)_{n \in \mathbb{N}}$ be a sequence of $\mathcal{L}^p(\mu)$-functions which converges to a measurable function f locally in measure. Then, the following are equivalent:*

(i) $\lim_{n \to \infty} \|f - f_n\|_p = 0$

(ii) $(|f_n|^p)_{n \in \mathbb{N}}$ *is uniformly integrable*

(iii) $\lim_{n \to \infty} \int |f_n|^p\, d\mu = \int |f|^p\, d\mu.$

Chapter 2

Finite Networks

2.1 Definition of Effective Resistance

Consider a network of resistors and two terminals x, y (e.g. Figure 2.1). If we attach a battery of voltage v_0 between x and y, an electric current of i_0 Ampere flows from x to y. Our goal is to replace the whole network "between" x and y with a single resistor R^* in an electronically equivalent fashion. More precisely, if we attach a battery of voltage v_0 to R^*, a current of i_0 Ampere should again flow from x to y. In this case, we call the resistance of R^* the network's **effective resistance** between x and y and denote it by $R(x, y)$.

A network of resistors is a weighted graph $G = (V, c)$ where each edge $(x, y) \in E(G)$ represents a resistor of **conductance** $c(x, y)$ and **resistance** $r(x, y) = c(x, y)^{-1}$. In this chapter we will only consider finite graphs.

Fix $x, y \in V$, $x \neq y$. An **electric current** from x to y transports electric charge (electrons) across the network. Hence, it is natural to model it as a **flow** on G, i.e. a function $I : E(G) \to \mathbb{R}$ satisfying

$$I(v, w) = -I(w, v) \tag{2.1}$$

for all $v, w \in V$. Then, $I(v, w)$ is the electric charge traveling directly along the edge (v, w). The total amount of current leaving x is

$$\sum_{v \in V} I(x, v) = c_x \cdot (\operatorname{div} I)(x). \tag{2.2}$$

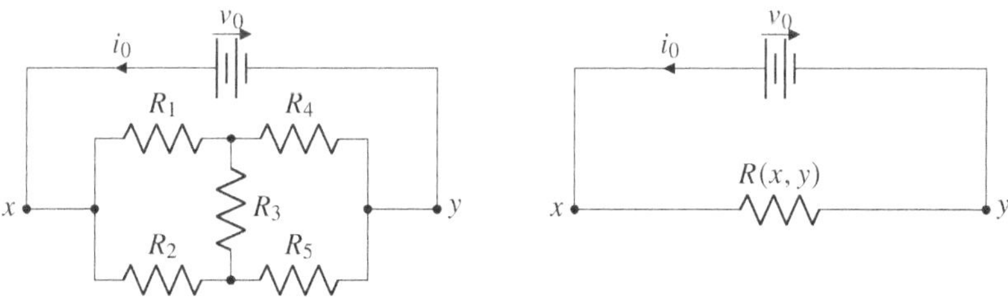

Figure 2.1: A network of resistors (left) and its electronically equivalent replacement (right)

In physics, there exist the three following laws describing properties of electric currents. The shown formulations appeared in [18].

Ohm's Law. The voltage across conducting materials is directly proportional to the current flowing through the material, or

$$V = R \cdot I, \tag{2.3}$$

where V is the potential difference in volts, R is the resistance in ohms and I is the current in amperes.

Kirchhoff's Voltage Law (KVL). The algebraic sum of the voltages around any closed path is zero.

Kirchhoff's Current Law (KCL). The algebraic sum of the currents entering any node [which is neither source nor sink] is zero.

By Ohm's Law, the voltage drop across a single resistor (xy) can be expressed as $r(x, y) \cdot I(x, y)$. Hence, in our model, Kirchhoff's Voltage Law becomes the following statement.

Postulate 1 (KVL). If I is an electric current, then

$$\sum_{k=0}^{n-1} r(x_k, x_{k+1}) \cdot I(x_k, x_{k+1}) = 0. \tag{2.4}$$

for every cycle $(x_0, \ldots, x_n)$ in G.

In our context, Kirchhoff's Current Law can be stated in the following way.

Postulate 2 (KCL). Let I be an electric current from x to y. Then, for $z \in V \setminus \{x, y\}$,

$$\sum_{v \in V} I(z, v) = 0. \tag{2.5}$$

Note that (2.5) is equivalent to $(\operatorname{div} I)(z) = 0$ for all $x \neq z \neq y$. Since a current I of i_0 Ampere from x to y satisfies $c_x \cdot (\operatorname{div} I)(x) = i_0$, Proposition 1.11 implies $c_y \cdot (\operatorname{div} I)(y) = -i_0$. Indeed, we have

$$c_x \cdot (\operatorname{div} I)(x) + c_y \cdot (\operatorname{div} I)(y) = \sum_{v \in V} c_v \cdot (\operatorname{div} I)(v) = \langle \mathbf{1}_V, \operatorname{div} I \rangle_V = \langle \nabla \mathbf{1}_V, I \rangle_E = 0. \tag{2.6}$$

Hence, (2.5) simply means that no electric charge appears or vanishes within our network except at the current's source x and sink y, respectively.

Consider our initial setup again and imagine that we exchanged our battery for a dynamic voltage generator. If we now adjust the applied voltage v_0 in such a way that exactly 1 Ampere flows from x to y, Ohm's Law states that the effective resistance between x and y is exactly v_0. Thus, for our purposes, it suffices to only consider currents of 1 Ampere and we will refer to them as ***unit currents***.

Corollary 2.1. *If I is a unit current from x to y, then there exists $\phi : V \to \mathbb{R}$ such that $I = \nabla \phi$ and*

$$\Delta \phi = \frac{1}{c_x} \mathbf{1}_x - \frac{1}{c_y} \mathbf{1}_y \tag{D}$$

$$\phi(y) = 0 \, .$$

Proof. For any path γ, let

$$I(\gamma) := \sum_{k=0}^{L(\gamma)-1} r(\gamma_k, \gamma_{k+1}) \cdot I(\gamma_k, \gamma_{k+1}).$$

For $\gamma^{-1} = (\gamma_{L(\gamma)}, \gamma_{L(\gamma)-1}, \ldots, \gamma_0)$, we then have $I(\gamma^{-1}) = -I(\gamma)$.

Define $\phi(y) := 0$ and let $v \in V, v \neq y$. Since G is connected, there exists a path $\gamma : v \to y$ and we define $\phi(v) := I(\gamma)$. KVL implies that this is independent of the choice of γ. Indeed, let γ' be another path $v \to y$. Denote by γ_c the concatenation of γ' and γ^{-1}. Then, γ_c is a cycle and by KVL, we have

$$0 = I(\gamma_c) = I(\gamma') + I(\gamma^{-1}) = I(\gamma') - I(\gamma).$$

Hence, $I(\gamma) = I(\gamma')$.

For $(v, w) \in E(G)$, let γ be a path $v \to y$. Then, (w, γ) is a path $w \to y$ and we have

$$(\nabla \phi)(v, w) = c(v, w)(\phi(v) - \phi(w)) = c(v, w)(I(\gamma) - I((w, \gamma)))$$
$$= c(v, w) \cdot r(w, v) \cdot (-I(w, v)) = I(v, w).$$

By KCL, we have

$$(\Delta \phi)(v) = (\operatorname{div} \nabla \phi)(v) = \operatorname{div} I(v) = \frac{1}{c_v} \sum_{w \in V} I(v, w) = \frac{1}{c_v} \cdot \begin{cases} 1 & , v = x \\ -1 & , v = y \\ 0 & , v \notin \{x, y\} \end{cases} \, .$$

$\square$

Proposition 2.2. *Let G be a possibly unconnected graph with no isolated vertices. Then, there exists a solution of (D) for every $x, y \in V$ if and only if G is connected. In this case, the solution is unique.*

Proof. Suppose that G is connected and let $x, y \in V, x \neq y$. We will show at a later point (Proposition 2.18) that the function

$$\phi(z) := \frac{1}{c_z} \mathbf{E}_x \left[\sum_{k=0}^{\tau_y - 1} \mathbf{1}_z(X_k) \right] .$$

solves (D). Since G is connected and finite, $(X_n)_{n \in \mathbb{N}}$ is irreducible and every state is positive recurrent. Hence, $\phi(z) \leq (c_z)^{-1} \cdot \mathbf{E}_x[\tau_y] < \infty$.

If ϕ_1, ϕ_2 are solutions to (D), then, $h := \phi_1 - \phi_2$ is harmonic. On a finite graph, we have $l^2(V) = l(V)$. Hence, by Proposition 1.12, h is constant. Since $h(y) = 0$, it follows that $\phi_1 = \phi_2$.

Now suppose that G is not connected and $X \subseteq V$ is a connected component of G. For all $v \in X$, we then have $c_v = \sum_{w \in X} c(v, w)$. Furthermore, $c(v, w) > 0$ implies $w \in X$. Hence, for any function $f : V \to \mathbb{R}$, we compute

$$\sum_{v \in X} (\Delta f)(v) \cdot c_v = \sum_{v \in X} \left(\sum_{w \in V} c(v, w) f(w) \right) - c_v f(v)$$
$$= \sum_{v \in X} \sum_{w \in X} c(v, w) f(w) - \sum_{v \in X} c_v f(v)$$
$$= \sum_{w \in X} f(w) c(v, X) - \sum_{v \in X} c_v f(v)$$
$$= \sum_{w \in X} c_w f(w) - \sum_{v \in X} c_v f(v) = 0.$$

Now let $x \in X$ and $y \in V \setminus X$. If $\phi : V \to \mathbb{R}$ satisfies $(\Delta \phi)(x) = 1/c_x$, then the above computation shows that there exists some $x' \in X$ such that $(\Delta \phi)(x') < 0$ and thus ϕ cannot solve $\Delta \phi = (c_x)^{-1} \mathbf{1}_x - (c_y)^{-1} \mathbf{1}_y$. Hence, there does not exist a solution for (D). $\square$

Ohm's Law states that if $I = \nabla \phi$ is a unit current from x to y, then the effective resistance between x and y equals the voltage drop $\phi(x) - \phi(y)$ across x and y. This motivates the following definition.

Definition 2.3. Let $x, y \in V$ and ϕ^{xy} be the unique solution of (D). Then,

$$R(x, y) := \phi^{xy}(x) \tag{2.7}$$

is the ***effective resistance*** of G between x and y. We call ϕ^{xy} the ***unit potential*** of G.

At times we may need to indicate which graph's effective resistance and potential we are considering in order to avoid confusion. In such instances we will write R_G and ϕ_G^{xy}, respectively.

Example 2.4 (A single edge). Consider the simplest graph possible consisting of only two vertices and one edge. Then, the effective resistance between x and y should be $R(x, y) = r(x, y)$. While this is clear in circuit theory, it presents an easy "sanity check" for our mathematical model. By definition, $\phi^{xy}(y) = 0$ and

$$\frac{1}{c_x} = (\Delta \phi^{xy})(x) = \phi^{xy}(x) - \frac{c(x, y)}{c_x} \cdot \phi^{xy}(y) = \phi^{xy}(x).$$

Hence, $R(x, y) = \phi^{xy}(x) = 1/c_x = c(x, y)^{-1} = r(x, y)$.

Example 2.5 (Unweighted complete graph). Let $n \in \mathbb{N}$ and consider the unweighted complete graph on n vertices. More precisely, let $V = \{1, \ldots, n\}$ and $G = (V, c)$ where $c(x, y) = 1 - \mathbf{1}_x(y)$. Note that $c_z = n - 1$ for all $z \in V$ and fix $x, y \in V, x \neq y$. By symmetry, we have $\phi^{xy}(z) = \phi^{xy}(z') =: k$ for all $z, z' \in V \setminus \{x, y\}$. Hence, (D) implies for $z \in \setminus \{x, y\}$

$$0 = (\Delta \phi^{xy})(z) = k - \frac{n-3}{n-1} \cdot k - \frac{1}{n-1} \cdot \phi^{xy}(x) = \frac{1}{n-1} (2k - \phi^{xy}(x))$$

and $\phi^{xy}(x) = 2k$ follows. By (D), we also have

$$\frac{1}{n-1} = (\Delta\phi^{xy})(x) = \phi^{xy}(x) - \frac{n-2}{n-1} \cdot k = 2k - \frac{n-2}{n-1} \cdot k = k \cdot \frac{n}{n-1}.$$

It follows that $\phi^{xy}(x) = 2/n$, $\phi^{xy}(y) = 0$ and $\phi^{xy}(z) = 1/n$ for all $z \in V \setminus \{x, y\}$. In particular,

$$R(x, y) = \frac{2}{n} \cdot (1 - \mathbf{1}_x(y)).$$

2.2 Network Reduction

While we can compute a network's effective resistance by solving the linear equation system (D) for every $x, y \in V$, there is an alternative method frequently used in circuit theory, namely **network reduction**. Its main idea is to repeatedly remove a single vertex (and all incident edges) while adjusting the remaining edge weights in such a way that the effective resistances of between all remaining vertices do not change. Once only two vertices x, y remain, the effective resistance $R(x, y)$ equals the pair-wise resistance $r(x, y)$ in the reduced network (see Example 2.4).

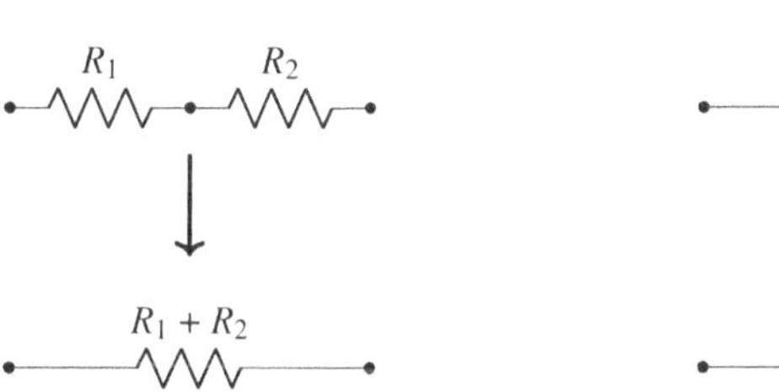

Figure 2.2: Replacing resistors connected in series.

Figure 2.3: Replacing resistors connected in parallel.

Widely known versions of this concept are rules about replacing resistors in series and parallel circuits. If two resistors with resistances R_1 and R_2 are connected in series, they can be replaced by a single resistor of resistance $R_1 + R_2$, see Figure 2.2. If the two are connected in parallel, the resistance of the replacing resistor must be $(R_1^{-1} + R_2^{-1})^{-1}$, see Figure 2.3.

The most general method of network reduction is the **Star-Mesh transform** or **Kron reduction**, see e.g. [4, 8]. As its name suggests, it transforms star-shaped circuits into mesh-shaped ones, see Figure 2.4, thereby removing the star's center vertex. In our language of weighted graphs, it can be formulated as follows.

Theorem 2.6 (Star-Mesh-Transform). *Let (V, c) be a finite graph with effective resistance R and fix $x^* \in V$. Furthermore, let R' denote the effective resistance of the graph $(V \setminus \{x^*\}, c')$ where*

$$c'(x, y) := c(x, y) + \frac{c(x, x^*) \cdot c(x^*, y)}{c_{x^*}}. \tag{2.8}$$

for all $x \neq y$. Then, $R'(x, y) = R(x, y)$ for all $x, y \in V \setminus \{x^*\}$.

Proof. Let $x, y \in V \setminus \{x^*\}$ and let ϕ^{xy} denote the potential of the unit current between x and y in (V, c), i.e. ϕ^{xy} is the unique solution of the linear equation system

$$c_x \phi^{xy}(x) - \sum_{w \neq x} c(x, w) \phi^{xy}(w) = 1$$

$$c_z \phi^{xy}(z) - \sum_{w \neq z} c(z, w) \phi^{xy}(z) = 0 \quad \forall z \neq x, y$$

$$\phi^{xy}(y) = 0$$

We now show that ϕ^{xy} solves the analogous equation system for the graph $(V \setminus \{x^*\}, c')$. First, note that for $z \neq x^*$, we have

$$c'_z = c_z - c(z, x^*) + \sum_{w \neq z, x^*} \frac{c(z, x^*) c(x^*, w)}{c_{x^*}}$$

$$= c_z + c(z, x^*) \left(-1 + \frac{c_{x^*} - c(x^*, z)}{c_{x^*}} \right) = c_z - \frac{c(x^*, z)^2}{c_{x^*}} \ .$$

Hence,

$$c'_x \phi^{xy}(x) - \sum_{w \neq x, x^*} c'(x, w) \phi^{xy}(w)$$

$$= c_x \phi^{xy}(x) - \sum_{w \neq x, x^*} c(x, w) \phi^{xy}(w) - \frac{c(x^*, x)^2}{c_{x^*}} \phi^{xy}(x) - \sum_{w \neq x, x^*} \frac{c(x, x^*) c(x^*, w)}{c_{x^*}} \phi^{xy}(w)$$

$$= 1 + c(x, x^*) \phi^{xy}(x^*) - \frac{c(x^*, x)^2}{c_{x^*}} \phi^{xy}(x) - \sum_{w \neq x, x^*} \frac{c(x, x^*) c(x^*, w)}{c_{x^*}} \phi^{xy}(w)$$

$$= 1 + c(x, x^*) \left(\phi^{xy}(x^*) - \sum_{w \neq x^*} \frac{c(x^*, w)}{c_{x^*}} \phi^{xy}(w) \right) = 1 \ .$$

Analogously, for $z \in V \setminus \{x, y, x^*\}$, we have

$$c'_z \phi^{xy}(z) - \sum_{w \neq z, x^*} c'(z, w) \phi^{xy}(z) = c_z \phi^{xy}(z) - \overbrace{\sum_{w \neq z, x^*} c(z, w) \phi^{xy}(w)}^{= c(z, x^*) \phi^{xy}(x^*)}$$

$$- \frac{c(x^*, z)^2}{c_{x^*}} \phi^{xy}(z) - \sum_{w \neq z, x^*} \frac{c(z, x^*) c(x^*, w)}{c_{x^*}} \phi^{xy}(w)$$

$$= c(z, z_0) \left(\phi^{xy}(x^*) - \sum_{w \neq x^*} \frac{c(x^*, w)}{c_{x^*}} \phi^{xy}(w) \right) = 0.$$

Since we also have $\phi^{xy}(y) = 0$, we see that ϕ^{xy} is indeed the potential of the unit current in $(V \setminus \{x^*\}, c')$ and thus $R'(x, y) = R(x, y)$ holds. $\qquad\square$

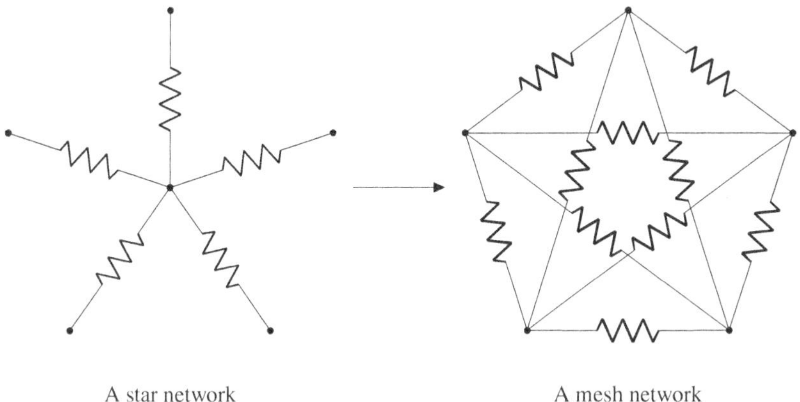

A star network   A mesh network

Figure 2.4: The Star-Mesh transform

Remark 2.7. As seen in the proof, (2.8) implies an explicit formula for the sum of edge weights at a given vertex x, namely

$$c'_x = c_x - \frac{c(x^*,x)^2}{c_{x^*}} . \tag{2.9}$$

We can also formulate (2.8) in terms of edge-resistances

$$r'(x,y) = \left(\frac{1}{r(x,y)} + \frac{1}{r(x,x^*) + r(x^*,y)} \right)^{-1} . \tag{2.10}$$

Remark 2.8 (Probabilistic interpretation)**.** Let $\mathbf{P}_x$ and $\mathbf{P}'_x$ denote the random walk on (V,c) and $(V \setminus \{x^*\}, c')$, respectively, and let $y \in V \setminus \{x^*, x\}$. Then,

$$
\begin{aligned}
\mathbf{P}'_x[X_1 = y] = \frac{c'(x,y)}{c'_x} &= \frac{c_x}{c'_x} \left(\frac{c(x,y)}{c_x} + \frac{c(x,x^*)c(x^*,y)}{c_x c_{x^*}} \right) \\
&= \left(1 - \frac{c(x,x^*)c(x^*,x)}{c_x c_{x^*}} \right)^{-1} (\mathbf{P}_x[X_1 = y] + \mathbf{P}_x[X_1 = x^*, X_2 = y]) \\
&= (\mathbf{P}_x[(X_1, X_2) \neq (x^*,x)])^{-1} \cdot (\mathbf{P}_x[X_1 = y] + \mathbf{P}_x[X_1 = x^*, X_2 = y]) \\
&= \mathbf{P}_x[X_1 = y \vee X_1 = x^*, X_2 = y \mid (X_1, X_2) \neq (x^*,x)].
\end{aligned}
$$

The behavior of $\mathbf{P}'_x$ is very similar to that of $\mathbf{P}_x$ with only two differences. First, the removal of x^* from (V,c) is compensated by adding a ***shortcut*** from x to y, enabling $\mathbf{P}'_x$ to jump directly from x to y whenever $\mathbf{P}_x$ would have taken a detour over x^*. Second, the transition probabilities of $\mathbf{P}'_x$ are those of $\mathbf{P}_x$ conditioned on $(X_1, X_2) \neq (x^*,x)$. This is due to the fact that our version of the star-mesh transform stays within our class of weighted graphs; more precisely, graphs which do not admit self-loops. Hence, the 'shorted' random walk forgets about the possibility of going to x^* and back to x again.

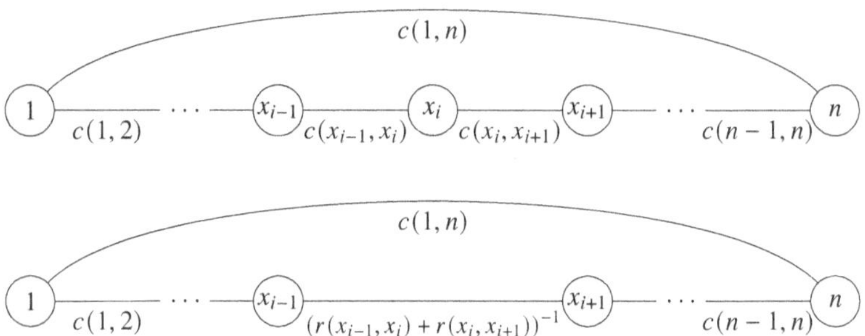

Figure 2.5: A cycle graph with n vertices (top) and the graph G' resulting from applying the Star-Mesh transform to x_i (bottom).

Remark 2.9. The Star-Mesh transform is also known as **Kron reduction** [8] and corresponds to computing the Schur complement of the Laplacian matrix.

Example 2.10. Let $n \in \mathbb{N}$, $n \geq 3$, and consider a graph which consists of a simple cycle on n vertices with arbitrary weights, i.e. $G = (\{1, \ldots, n\}, c)$ with $c(x, y) > 0$ if and only if $|x - y| = 1$ or $\{x, y\} = \{1, n\}$, see Figure 2.5. For $x, y \in V$, there are exactly two simple paths γ_1, γ_2 in G from x to y. We claim that

$$R(x, y) = \left(\frac{1}{r(\gamma_1)} + \frac{1}{r(\gamma_2)} \right)^{-1}$$

and will prove it using the Star-Mesh transform and induction over n.

For $n = 3$, let $\{x, y, z\} = \{1, 2, 3\}$. Then, $\gamma_1 = (x, y)$ and $\gamma_2 = (x, z, y)$. Applying the Star-Mesh transform to z, we have

$$R(x, y) = r'(x, y) = \left(c(x, y) + \frac{c(x, z) \cdot c(z, y)}{c(z, x) + c(z, y)} \right)^{-1}$$

$$= \left(\frac{1}{r(x, y)} + \frac{1}{r(x, z) + r(z, y)} \right)^{-1} = \left(\frac{1}{r(\gamma_1)} + \frac{1}{r(\gamma_2)} \right)^{-1}.$$

For $n > 3$, let $x, y \in V$. Without loss of generality, $\gamma_1 = (x_0, \ldots, x_m)$ satisfies $m > 1$. Choose $i \in \{1, \ldots, m - 1\}$ and apply the Star-Mesh transform to x_i. This yields a graph $G' = (V', c')$ where $V' = \{1, \ldots, n\} \setminus \{x_i\}$. Since $c(x_i, v) = 0$ for all $v \in V \setminus \{x_{i-1}, x_{i+1}\}$, we have $c(v, w) = c'(v, w)$ for all $v, w \in V' \setminus \{x_{i-1}, x_{i+1}\}$ and

$$c'(x_{i-1}, x_{i+1}) = \frac{c(x_{i-1}, x_i) \cdot c(x_i, x_{i+1})}{c(x_{i-1}, x_i) + c(x_i, x_{i+1})} = \frac{1}{r(x_{i-1}, x_i) + r(x_i, x_{i+1})}$$

as depicted in Figure 2.5. The graph G' is itself a cycle with $n - 1$ vertices. Let $\gamma_1' = (x_0, \ldots, x_{i-1}, x_{i+1}, \ldots, x_m)$ and $\gamma_2' = \gamma_2$ be the two distinct simple paths $x \to y$ in G'. Then,

$$r(\gamma_1') = \sum_{k=0}^{i-2} r(x_k, x_{k+1}) + r(x_{i-1}, x_i) + r(x_i, x_{i+1}) + \sum_{k=i+1}^{m-1} r(x_k, x_{k+1}) = r(\gamma_1)$$

and, by induction,

$$R(x, y) = \left(\frac{1}{r(\gamma_1')} + \frac{1}{r(\gamma_2')} \right)^{-1} = \left(\frac{1}{r(\gamma_1)} + \frac{1}{r(\gamma_2)} \right)^{-1}.$$

$\square$

Lemma 2.11. *Let $z \in V$ and $G' = (V \setminus \{z\}, c')$ be the graph that results from applying the Star-Mesh transform to z. For any path $\gamma' : x \to y$ in G', there exists a path $\gamma : x \to y$ in G such that γ' results from removing all occurrences of z from γ.*

Proof. Let $\gamma' = (x_0, \dots, x_n)$. We use induction on n to prove the claim.
    Let $n = 1$, i.e. $\gamma' = (x, y)$ Then,

$$c'(x, y) = c(x, y) + \frac{c(x, z) \cdot c(z, y)}{c_z} > 0.$$

If $c(x, y) > 0$, then (x, y) is a path in G and we are done. If $c(x, y) = 0$, it follows that $c(x, z)c(z, y) > 0$ and therefore $\gamma = (x, z, y)$ is a path in G. Removing z from γ then yields γ'.
    Let $n > 1$ and $p' = (x_0, \dots, x_{n-1})$. By induction, there exists a path $p : x_0 \to x_{n-1}$ in G such that removing all occurrences of z from p yields p'. Since γ' is a path in G', we have

$$c'(x_{n-1}, y) = c(x_{n-1}, y) + \frac{c(x_{n-1}, z) \cdot c(z, y)}{c_z} > 0.$$

If $c(x_{n-1}, y) > 0$, then, $\gamma = (p, y)$ is a path in G and removing all occurrences of z from it yields $(p', y) = \gamma'$. If $c(x_{n-1}, y) = 0$, then $c(x_{n-1}, z) \cdot c(z, y)$ follows and $\gamma = (p, z, y)$ is a path in G with the desired property. $\square$

Let $W \subseteq V$. We say a path $\gamma = (x_0, \dots, x_n)$ is ***outside of*** W if $x_1, \dots, x_{n-1} \notin W$, i.e. every vertex except x_1, x_{n-1} lies outside of W.

Proposition 2.12. *Let $W \subseteq V$. If for all $x, y \in W$, there exists no path $x \to y$ outside of W, then $R_G(x, y) = R_{G \restriction W}(x, y)$ for all $x, y \in W$.*

Proof. Let $G = (V, c)$ and $V \setminus W = \{v_1, \dots, v_n\}$. For $m = 0, \dots, n$, let $V_m := V \setminus \{v_1, \dots, v_m\}$ and $G_m = (V_m, c_m)$ be the graph that results from iteratively applying the Star-Mesh transform (Theorem 2.6) to $v_1, \dots, v_m$. Using induction on $m = 0, \dots, n$, we will show that, for all $x, y \in W$,

$$c_m(x, y) = c(x, y)$$

holds and that there exists no path $x \to_{G_m} y$. In particular, this means that $G \restriction W = G \restriction V_n = G_n$ and

$$R_G(x, y) = R_{G \restriction W}(x, y) \ \forall \, x, y \in W$$

follows.

For $m = 0$, there is nothing to show since $G_0 = G$. For $m > 0$, G_m is obtained by applying the Star-Mesh transform to v_m in G_{m-1}. For $x, y \in W$, we have

$$c_m(x, y) = c_{m-1}(x, y) + \frac{c_{m-1}(x, v_m) \cdot c_{m-1}(v_m, y)}{(c_{m-1})_{v_m}}.$$

By induction, there exists no path $x \to_{G_{m-1}} y$ outside of W and it follows that $c_{m-1}(x, v_m) = 0$ or $c_{m-1}(v_m, y) = 0$. Hence, $c_m(x, y) = c_{m-1}(x, y) = c(x, y)$ where the last equality is also due to induction.

Now suppose that γ_m is a path $x \to_{G_m} y$ outside of W. Then, by Lemma 2.11, there exists a path $\gamma_{m-1} : x \to y$ in G_{m-1} such that removing all occurrences of v_m in γ_{m-1} yields γ_m. Since γ_m lies outside of W and $v_m \notin W$, γ_{m-1} also lies outside of W. By induction, however, there exists no such path in G_{m-1}.

$\square$

2.3　Resistance by Optimization

It is long known that electric currents are flows which minimize energy dissipation [10, 16]. This is often referred to as Thomson's principle and can be related to the nature of the current's potential. It enables us to compute the effective resistance by solving variational problems related to the energy form.

Theorem 2.13 (Thomson's Principle [38]). *Let $I = \nabla\phi^{xy}$ be the unit current from x to y. Then,*

$$\mathcal{E}(I) = \min\left\{\mathcal{E}(J) \mid J \in l^2_-(E), \operatorname{div} J = \frac{1}{c_x}\mathbf{1}_x - \frac{1}{c_y}\mathbf{1}_y\right\}. \tag{2.11}$$

Proof. Since $\operatorname{div} I = \operatorname{div} J$, we have

$$\begin{aligned}
\mathcal{E}(J) &= \mathcal{E}(I) + 2\langle I, J - I\rangle_E + \mathcal{E}(J - I) \\
&= \mathcal{E}(I) + 2\langle \phi^{xy}, \operatorname{div}(J - I)\rangle_V + \mathcal{E}(J - I) \\
&= \mathcal{E}(I) + \mathcal{E}(J - I) \geq \mathcal{E}(I).
\end{aligned}$$

$\square$

By Proposition 1.11, we have

$$\mathcal{E}(\phi^{xy}) = \langle \Delta\phi^{xy}, \phi^{xy}\rangle_V = \phi^{xy}(x) - \phi^{xy}(y) = R(x, y). \tag{2.12}$$

Proposition 2.14. *For $x \neq y$, let $\psi^{xy} := \phi^{xy}/\mathcal{E}(\phi^{xy})$. Then, ψ^{xy} is the unique minimizer of $\mathcal{E}$ among all $f \in l(V)$ with $f(x) = 1$ and $f(y) = 0$.*

Proof. First, note that $\psi^{xy}(x) = \phi^{xy}(x)/\mathcal{E}(\phi^{xy}) = 1$ and $\psi^{xy}(y) = 0$.

Let $f \in l(V)$, $f(x) = 1$, $f(y) = 0$. Since $\mathcal{E}$ is bilinear, we have

$$
\begin{aligned}
\mathcal{E}(f) &= \mathcal{E}(\psi^{xy} + (f - \psi^{xy})) \\
&= \mathcal{E}(\psi^{xy}) + 2\mathcal{E}(\psi^{xy}, f - \psi^{xy}) + \mathcal{E}(f - \psi^{xy}) \\
&= \mathcal{E}(\psi^{xy}) + 2\langle \Delta\psi^{xy}, f - \psi^{xy}\rangle_V + \mathcal{E}(f - \psi^{xy}) \\
&= \mathcal{E}(\psi^{xy}) + \frac{2}{\mathcal{E}(\phi^{xy})} \underbrace{\langle \Delta\phi^{xy}, f - \psi^{xy}\rangle_V}_{=0} + \mathcal{E}(f - \psi^{xy}) \\
&= \mathcal{E}(\psi^{xy}) + \mathcal{E}(f - \psi^{xy}) \geq \mathcal{E}(\psi^{xy}).
\end{aligned}
$$

Hence,

$$
\mathcal{E}(\psi^{xy}) = \min\{\mathcal{E}(f) \mid f \in l(V), f(x) = 1, f(y) = 0\}
$$

Furthermore, if $\mathcal{E}(f) = \mathcal{E}(\psi^{xy})$, then $\mathcal{E}(f - \psi^{xy}) = 0$ and it follows that $f - \psi^{xy}$ is constant. Since $f(x) = \psi^{xy}(x) = 1$, $f = \psi^{xy}$ follows. $\qquad\square$

Computing

$$
\mathcal{E}(\psi^{xy}) = \mathcal{E}\left(\frac{\phi^{xy}}{\mathcal{E}(\phi^{xy})}\right) = \mathcal{E}(\phi^{xy})^{-1} = R(x, y)^{-1},
$$

we get the following Corollary which describes the effective resistance as the solution to a variational problem.

Corollary 2.15. *For $x, y \in V$, $x \neq y$, we have*

$$
R(x, y) = (\min\{\mathcal{E}(f) \mid f \in l(V), f(x) = 1, f(y) = 0\})^{-1}. \tag{2.13}
$$

Sometimes, it will be more useful to consider an alternative variational problem which does not restrict the values at x and y but also does not have a unique solution.

Corollary 2.16. *For $x, y \subset V$, $x \neq y$, we have*

$$
R(x, y) = \max\left\{\frac{(f(x) - f(y))^2}{\mathcal{E}(f)} \mid f \in l(V), \mathcal{E}(f) > 0\right\}. \tag{2.14}
$$

Proof. For any $f \in l(V)$ with $f(x) \neq f(y)$, let

$$
\tilde{f}(v) := \frac{f(v) - f(y)}{f(x) - f(y)}.
$$

Then, $\tilde{f}(x) = 1$, $\tilde{f}(y) = 0$ and we have

$$
\frac{(f(x) - f(y))^2}{\mathcal{E}(f)} = \mathcal{E}(\tilde{f})^{-1} \leq \mathcal{E}(\psi^{xy})^{-1} = \frac{(\psi^{xy}(x) - \psi^{xy}(y))^2}{\mathcal{E}(\psi^{xy})} = R(x, y).
$$

$\qquad\square$

It follows immediately that

$$
(f(x) - f(y))^2 \leq R(x, y) \cdot \mathcal{E}(f) \tag{2.15}
$$

for all $f \in l(V)$.

2.4 Probabilistic Representations

We use the approach detailed in [37] and apply it to our context of weighted graphs to establish a direct connection between the graph's effective resistance and random walk.

Definition 2.17. For $x, y \in V$, let

$$\Phi^{xy} := \sum_{k=0}^{\tau_y - 1} \mathbf{1}_x(X_k)$$

be the number of times the random walk visits x before reaching y. For $\tau_y = 0$, we define $\Phi^{xy} = 0$.

Proposition 2.18. *For $x, y, z \in V$, we have*

$$\phi^{xy}(z) = \frac{1}{c_z} \mathbf{E}_x\left[\Phi^{zy}\right]. \tag{2.16}$$

In particular,

$$R(x, y) = \frac{1}{c_x} \mathbf{E}_x\left[\Phi^{xy}\right]. \tag{2.17}$$

Proof. For $z \in V$, let

$$\phi(z) := \frac{1}{c_z} \mathbf{E}_x[\Phi^{zy}].$$

We will show that ϕ solves (D). The first claim then follows by uniqueness of ϕ^{xy} while the second is due to $R(x, y) = \phi^{xy}(x)$.

First note that for $x = y$, $\Phi^{zy} \equiv 0$ holds $\mathbf{P}_x$-almost surely. Hence, $\phi \equiv 0$ and it solves (D).

For $x \neq y$, we have $\phi(y) = 0$ because $\Phi^{yy} \equiv 0$. For $z \neq y$, we use (MP) to compute

$$\mathbf{E}_x[\Phi^{zy}] = \mathbf{1}_x(z) + \sum_{k=1}^{\infty} \mathbf{P}_x[X_k = z, k \leq \tau_y] = \mathbf{1}_x(z) + \sum_{k=1}^{\infty} \mathbf{P}_x[X_k = z, k - 1 < \tau_y]$$

$$= \mathbf{1}_x(z) + \sum_{k=1}^{\infty} \sum_{v \in V\setminus\{y\}} \mathbf{P}_x[X_k = z \mid X_{k-1} = v, k - 1 < \tau_y] \cdot \mathbf{P}_x[X_{k-1} = v, k - 1 < \tau_y]$$

$$= \mathbf{1}_x(z) + \sum_{v \in V\setminus\{y\}} \left(\frac{c(v, z)}{c_v} \sum_{k=0}^{\infty} \mathbf{P}_x[X_{k-1} = v, k - 1 < \tau_y] \right)$$

$$= \mathbf{1}_x(z) + \sum_{v \in V\setminus\{y\}} \left(\frac{c(v, z)}{c_v} \mathbf{E}_x[\Phi^{vy}] \right).$$

Hence,

$$\phi(z) = \frac{1}{c_z} \mathbf{1}_x(z) + \sum_{v \in V\setminus\{y\}} \frac{c(v, z)}{c_z c_v} \mathbf{E}_x[\Phi^{vy}] = \frac{1}{c_x} \mathbf{1}_x(z) + \sum_{v \in V} \frac{c(z, v)}{c_z} \phi(v)$$

and it follows that

$$(\Delta\phi)(z) = \frac{1}{c_x}\mathbf{1}_x(z) \; \forall \, z \neq y.$$

To compute $(\Delta\phi)(y)$, use that since G is finite, we have $\phi \in l^2(V)$ and thus

$$0 = \langle \Delta\mathbf{1}, \phi \rangle_V = \langle \mathbf{1}, \Delta\phi \rangle_V = \sum_{v \in V} c_v (\Delta\phi)(v) = 1 + c_y(\Delta\phi)(y).$$

This concludes the proof. $\qquad\square$

Corollary 2.19.

$$\phi^{xy}(z) = \frac{\mathbf{P}_x[\tau_z < \tau_y]}{c_z \cdot \mathbf{P}_z[\tau_y \leq \tau_z^+]} . \tag{2.18}$$

In particular, we have

$$R(x, y) = \frac{1}{c_x \mathbf{P}_x[\tau_y \leq \tau_x^+]} = \frac{1}{c_x \mathbf{P}_x[\tau_y < \tau_x^+]} . \tag{2.19}$$

Proof. For $y = z$, we have $\mathbf{P}_x[\tau_z < \tau_y] = 0$ and $\mathbf{P}_z[\tau_y \leq \tau_z^+] = 1$. Hence, the right-hand side of (2.18) vanishes for $z = y$.

From now on assume that $y \neq z$. We have $\mathbf{P}_x[\Phi^{zy} = 0] = \mathbf{P}_x[\tau_y \leq \tau_z]$ and, for $k > 0$,

$$\mathbf{P}_x[\Phi^{zy} = k] = \mathbf{P}_x[\tau_z < \tau_y] \cdot \mathbf{P}_z[\Phi^{zy} = k].$$

For $p = \mathbf{P}_z[\tau_z^+ < \tau_y]$ and $k \in \mathbb{N}^+$, we can compute that

$$\mathbf{P}_z[\Phi^{zy} = k] = (1 - p) \cdot p^{k-1}$$

by virtue of the random walk's Markov property. It follows that

$$\begin{aligned}
\mathbf{E}_x[\Phi^{zy}] &= \sum_{k=0}^{\infty} k \cdot \mathbf{P}_x[\Phi^{zy} = k] = \mathbf{P}_x[\tau_z < \tau_y] \cdot \sum_{k=1}^{\infty} k(1-p)p^{k-1} \\
&= \frac{\mathbf{P}_x[\tau_z < \tau_y]}{1 - p} = \frac{\mathbf{P}_x[\tau_z < \tau_y]}{\mathbf{P}_z[\tau_y \leq \tau_z^+]} .
\end{aligned}$$

The claim follows by (2.16). $\qquad\square$

Corollary 2.20. *For $x \neq y$, we have*

$$c(x, y) \leq \frac{1}{R(x, y)} \leq c_x. \tag{2.20}$$

Proof. By (2.19), we have

$$\frac{1}{R(x, y)} = c_x \cdot \mathbf{P}_x[\tau_y < \tau_x^+] \geq c_x \cdot \mathbf{P}_x[X_1 = y] = c_x \cdot \frac{c(x, y)}{c_x} = c(x, y).$$

The second inequality follows directly from (2.17) and the fact that $\mathbf{E}_x[\Phi^{xy}] \geq 1$. $\qquad\square$

Proposition 2.21. *For $x, y, z \in V$, we have $\phi^{xy}(z) = \phi^{zy}(x)$.*

Proof. Let $x, y \in V$. If $x = y$, then $\phi^{xy}(z) = \phi^{yy}(z) = 0 = \phi^{zy}(y) = \phi^{zy}(x)$ for all $z \in V$.

Now assume that $x \neq y$ and let $f(z) := \phi^{zy}(x)$ for all $z \in V$. Then, we have $f(y) = \phi^{yy}(x) = 0$ and, for $z \neq y$, we apply (MP) to do a computation similar to the proof of Proposition 2.18.

$$
\begin{aligned}
\mathbf{E}_z[\Phi^{xy}] &= \mathbf{1}_x(z) + \sum_{k=1}^{\infty} \mathbf{P}_z[X_k = x, k < \tau_y] \\
&= \mathbf{1}_x(z) + \sum_{k=1}^{\infty} \sum_{v \in V \setminus \{y\}} \mathbf{P}_z[X_k = x, k < \tau_y \mid X_1 = v] \cdot \mathbf{P}_z[X_1 = v] \\
&= \mathbf{1}_x(z) + \sum_{v \in V \setminus \{y\}} \left(\frac{c(z, v)}{c_z} \cdot \sum_{k=1}^{\infty} \mathbf{P}_v[X_{k-1} = x, k - 1 < \tau_y] \right) \\
&= \mathbf{1}_x(z) + \sum_{v \in V \setminus \{y\}} \frac{c(z, v)}{c_z} \cdot \mathbf{E}_v[\Phi^{xy}]
\end{aligned}
$$

Hence,

$$
\begin{aligned}
f(z) = \phi^{zy}(x) = \frac{1}{c_x} \mathbf{E}_z[\Phi^{xy}] &= \frac{1}{c_x} \mathbf{1}_x(z) + \sum_{v \in V \setminus \{y\}} \frac{c(z, v)}{c_z} \cdot \frac{1}{c_x} \mathbf{E}_v[\Phi^{xy}] \\
&= \frac{1}{c_x} \mathbf{1}_x(z) + \sum_{v \in V} \frac{c(z, v)}{c_z} f(v)
\end{aligned}
$$

and it follows that

$$
(\Delta f)(z) = \frac{1}{c_x} \mathbf{1}_x(z)
$$

for $z \neq y$. Using the same argument as at the end of the proof of Proposition 2.18, $(\Delta f)(y) = -1/c_y$ follows as well. Hence, f solves (D) which has the unique solution ϕ^{xy}. Hence, $\phi^{xy}(z) = f(z) = \phi^{zy}(x)$ follows for all $z \in V$. $\qquad\square$

Remark 2.22. We have just reproduced a special case of the ***Reciprocity Theorem*** from Circuit Theory which [18] states as follows.

> In any passive linear bilateral network, if the single current source I_x between nodes x and x' produces the voltage response V_y between nodes y and y', then the removal of the current source from nodes x and x' and its insertion between nodes y and y' will produce the voltage response V_y between nodes x and x'.

In our context, this more general statement corresponds to

$$
\phi^{xx'}(y) - \phi^{xx'}(y') = \phi^{yy'}(x) - \phi^{yy'}(x') \tag{2.21}
$$

for all $x, x', y, y' \in V$.

2.5 Metric Properties

A **metric space** is a pair (X, d) consisting of a non-empty set X and a non-negative function $d : X \times X \to [0, +\infty)$ satisfying

1. $d(x, y) = 0 \Leftrightarrow x = y$ for all $x, y \in X$,

2. $d(x, y) = d(y, x)$ for all $x, y \in X$ and

3. $d(x, y) \le d(x, z) + d(z, y)$ for all $x, y, z \in X$.

The last condition is called **triangle inequality**. We will now investigate the metric properties of the effective resistance.

For $x = y$, the Dirichlet Problem (D) becomes

$$\Delta\phi = 0$$
$$\phi(y) = 0$$

which, by Proposition 1.12, has only $\phi \equiv 0$ as solution. Hence,

$$R(x, x) = 0 \tag{2.22}$$

follows. For $x \ne y$, let $\phi := \phi^{xy}(x) - \phi^{xy}$. Then, $\phi(x) = 0$ and $\Delta\phi = (c_y)^{-1}\mathbf{1}_y - (c_x)^{-1}\mathbf{1}_x$. Since (D) has exactly one solution, $\phi^{yx} = \phi$ follows. Hence, we have

$$R(x, y) = \phi^{xy}(z) + \phi^{yx}(z) \tag{2.23}$$

for all $z \in V$ and, in particular,

$$R(x, y) = R(y, x). \tag{2.24}$$

A precise version of the triangle inequality was proved by Tetali in [37] for the unweighted case. We adopt the statement to our notion of weighted graphs.

Proposition 2.23 (Tetali).

$$\frac{1}{2}\left(R(x, y) + R(y, z) - R(x, z)\right) = \phi^{xy}(z). \tag{2.25}$$

Proof. Using (2.23) and Proposition 2.21, we have

$$R(x, y) + R(y, z) - R(x, z) = \phi^{xy}(z) + \phi^{yx}(z) + \phi^{yz}(x) + \phi^{zy}(x) - \phi^{xz}(y) - \phi^{zx}(y)$$
$$= \phi^{xy}(z) + \phi^{zy}(x) = 2 \cdot \phi^{xy}(z).$$

$$\square$$

Theorem 2.24 (Tetali). *The effective resistance R is a metric on V.*

Proof. We have $R(x, x) = 0$ and $R(x, y) = R(y, x)$ for all $x, y \in V$ by (2.22) and (2.24). By Proposition 2.23 and 2.16, we have

$$R(x, y) + R(y, z) - R(x, z) = 2 \cdot \phi^{xy}(z) = \frac{2}{c_z} \mathbf{E}_x[\Phi^{zy}] \geq 0$$

which implies the triangle inequality

$$R(x, y) + R(y, z) \geq R(x, z).$$

$\square$

Definition 2.25. We say that a finite metric space (V, d) is an ***effective resistance space (ERS)*** if there exists a graph $G = (V, c)$ with effective resistance d.

Proposition 2.26. *If (V, d) is an ERS and $\emptyset \neq W \subseteq V$, then $(W, d\restriction_W)$ is also an ERS.*

Proof. The claim follows by applying the Star-Mesh transform successively to all vertices in $V \setminus W$. $\square$

Proposition 2.27. *Let $x, y, z \in V$ be pairwise distinct vertices. Then,*

$$R(x, y) + R(y, z) - R(x, z) = 0$$

if and only if all paths $x \to z$ in G use y.

Proof. Let $x, y, z \in V$ be pairwise distinct and let $(v_0, \ldots, v_n)$ be any path $x \to z$. Then,

$$\mathbf{E}_x\left[\sum_{k=0}^{\tau_y - 1} \mathbf{1}_z(X_k)\right] = \sum_{k=0}^{\infty} \mathbf{P}_x\left[X_k = z, k < \tau_y\right]$$

$$\geq \mathbf{P}_x\left[X_n = z, n < \tau_y\right]$$

$$\geq \mathbf{P}_x[X_0 = x, X_1 = v_1, \ldots, X_{n-1} = v_{n-1}, X_n = z, n < \tau_y] \geq 0.$$

If $R(x, y) + R(y, z) - R(x, z) = 0$, then it follows by (2.25) and (2.16) that

$$\mathbf{P}_x[X_0 = x, X_1 = v_1, \ldots, X_{n-1} = v_{n-1}, X_n = z, n < \tau_y] = 0.$$

Hence, there exists no path $x \to_G z$ which does not use y.

Now assume that all paths $x \to z$ use y. Hence, $\mathbf{P}_x[X_k = z, k < \tau_y] = 0$ for all $k \geq 0$. It follows that

$$\frac{c_z}{2}(R(x, y) + R(y, z) - R(x, z)) = \mathbf{E}_x\left[\sum_{k=0}^{\tau_y - 1} \mathbf{1}_z(X_k)\right] = \sum_{k=0}^{\infty} \mathbf{P}_x\left[X_k = z, k < \tau_y\right] = 0.$$

$\square$

2.5.1 Comparison with Geodesic Metric

Recall that

$$r(\gamma) := \sum_{k=0}^{n-1} r(\gamma_k, \gamma_{k+1}) \tag{2.26}$$

is the weighted length of a path $\gamma = (\gamma_0, \ldots, \gamma_n)$. For $x, y \in V$, let

$$d_G(x, y) := \inf \{r(\gamma) \mid \gamma \in \Gamma_G(x, y)\} \tag{2.27}$$

be the **geodesic metric of** G. We say $\gamma : x \to y$ is **a shortest path between x and** y if $r(\gamma) = d_G(x, y)$. The fact that d_G actually defines a metric on V is a standard argument and we will leave the proof as an exercise.

When a metric on the vertices of a graph is needed, d_G is often a canonical choice. In the following example, we will see that R captures more data about G than d_G does.

Example 2.28. Consider the two graphs shown in Figure 2.6. Both have the same geodesic metric, namely

$$(d_G(v_i, v_j))_{i,j \in \{0,\ldots,3\}} = \begin{pmatrix} 0 & 1 & 2 & 1 \\ 1 & 0 & 1 & 2 \\ 2 & 1 & 0 & 1 \\ 1 & 2 & 1 & 0 \end{pmatrix}.$$

In fact, there are infinitely many graphs with this geodesic metric. Indeed, as long as $c(v_1, v_3) \in [0, 1/2]$, the geodesic metric does not change.

Now lets consider the effective resistances. The graph on the left has

$$(R_l(v_i, v_j))_{i,j \in \{0,\ldots,3\}} = \begin{pmatrix} 0 & 3/4 & 1 & 3/4 \\ 3/4 & 0 & 3/4 & 1 \\ 1 & 3/4 & 0 & 3/4 \\ 3/4 & 1 & 3/4 & 0 \end{pmatrix}$$

while the effective resistance of the graph on the right is given by

$$(R_r(v_i, v_j))_{i,j \in \{0,\ldots,3\}} = \begin{pmatrix} 0 & 2/3 & 1 & 2/3 \\ 2/3 & 0 & 2/3 & 2/3 \\ 1 & 2/3 & 0 & 2/3 \\ 2/3 & 2/3 & 2/3 & 0 \end{pmatrix}.$$

Hence, we are able to differentiate between the given graphs only by looking at their effective resistance. In the next section we will extend this to show that we can fully reconstruct the graph when given its effective resistance.

Lemma 2.29. *For $f \in l(V)$, $x, y \in V$, we have*

$$(f(x) - f(y))^2 \le d_G(x, y) \cdot \mathcal{E}(f). \tag{2.28}$$

In particular, $R(x, y) \le d_G(x, y)$ for all $x, y \in V$.

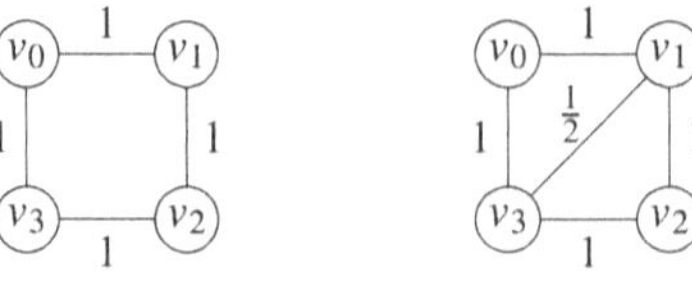

Figure 2.6:   Two graphs with the same geodesic metric.

Proof. Let $f \in l(V)$, $x, y \in V$ and $\gamma = (x_0, \dots, x_n)$ a shortest path $x \to y$, i.e. $r(\gamma) = d_G(x, y)$. Using the Cauchy-Schwarz inequality, we compute

$$
\begin{aligned}
(f(x) - f(y))^2 &= \left(\sum_{k=0}^{n-1} f(x_{k+1}) - f(x_k) \right)^2 \\
&= \left(\sum_{k=0}^{n-1} \frac{1}{\sqrt{c(x_{k+1}, x_k)}} \cdot \sqrt{c(x_{k+1}, x_k)} \cdot (f(x_{k+1}) - f(x_k)) \right)^2 \\
&\leq \left(\sum_{k=0}^{n-1} \frac{1}{c(x_{k+1}, x_k)} \right) \cdot \left(\sum_{k=0}^{n-1} c(x_{k+1}, x_k) \cdot (f(x_{k+1}) - f(x_k))^2 \right) \\
&\leq r(\gamma) \cdot \mathcal{E}(f) = d_G(x, y) \cdot \mathcal{E}(f).
\end{aligned}
$$

In particular,

$$
R(x, y) = \frac{(\phi^{xy}(x) - \phi^{xy}(y))^2}{\mathcal{E}(\phi^{xy})} \leq d_G(x, y).
$$

$\square$

Proposition 2.30. *On any tree G, $R = d_G$ holds.*

Proof. Let $x, y \in V$, $x \neq y$. Since $G = (V, c)$ is a tree, there exists a unique simple path $\gamma = (x_0, \dots, x_n)$ from x to y. For $i \in \{0, \dots, n\}$, define

$$
\phi(x_i) := d_G(x_i, y).
$$

For $z \in V \setminus \{x_0, \dots, x_n\}$, there exists a unique path γ_z from z to y. Since γ and γ_z both visit y, there exists $i(z) \in \{0, \dots, n\}$ such that $x_{i(z)}$ is the first vertex shared by the paths γ and γ_z. Now, define $\phi(z) := \phi(x_{i(z)})$.

Let $z \in V \setminus \{x_0, \dots, x_n\}$, $z' \in V$ such that $c(z, z') > 0$. Then, γ_z and $\widetilde{\gamma}_z = (z, \gamma_{z'})$ are two paths in G from z to y. Since there exists only one simple path $z \to y$, it follows that $\widetilde{\gamma}_z$ is not simple or $\widetilde{\gamma}_z = \gamma_z$. If $\widetilde{\gamma}_z$ is not simple, it must visit z twice (since $\gamma_{z'}$ is simple) and $\gamma_{z'} = (z', \gamma_z)$ follows. In both cases, we have $x_{i(z)} = x_{i(z')}$ and thus $\phi(z) = \phi(z')$. Therefore, $c(z, z') \cdot (\phi(z) - \phi(z'))^2 = 0$ whenever z or z' are not in $\{x_0, \dots, x_n\}$.

Hence,

$$\mathcal{E}(\phi) = \frac{1}{2} \sum_{v,w \in V} c(v,w)(\phi(v) - \phi(w))^2 = \sum_{k=0}^{n-1} c(x_k, x_{k+1}) \cdot (\phi(x_k) - \phi(x_{k+1}))^2$$

$$= \sum_{k=0}^{n-1} c(x_k, x_{k+1}) \cdot (d_G(x_k, y) - d_G(x_{k+1}, y))^2 = \sum_{k=0}^{n-1} c(x_k, x_{k+1}) \cdot r(x_k, x_{k+1})^2$$

$$= \sum_{k=0}^{n-1} r(x_k, x_{k+1}) = r(\gamma) = d_G(x, y)$$

It follows that

$$d_G(x, y) \geq R(x, y) \geq \frac{(\phi(x) - \phi(y))^2}{\mathcal{E}(\phi)} = \frac{d_G(x, y)^2}{d_G(x, y)} = d_G(x, y)$$

and thus $R(x, y) = d_G(x, y)$. $\qquad\qquad\qquad\qquad\qquad\qquad\qquad\qquad\square$

Remark 2.31. Note that the converse statement also holds. See [19] for a sketch of the proof.

2.6   Reconstructing G from R

Suppose that we are given a finite metric space (V, d) and the information that d is the effective resistance of some unspecified graph G. We will show that it is possible to reconstruct G from its effective resistance.

If (V, d) is an ERS, Proposition 2.23 implies that we can obtain the electrical potential ϕ^{xy} of a unit current between two points $x, y \in V$ via

$$\phi^{xy}(z) = \frac{1}{2}(d(x, y) + d(y, z) - d(x, z)). \tag{2.29}$$

We can use this formula as a definition for arbitrary finite metric spaces. This gives us the ability to interpret the Dirichlet problem (D) as an equation system of the unknown variable c rather than unknown function ϕ. By grouping together certain equations, we obtain a family of equation systems whose solution, if it exists and satisfies an additional condition, is a graph with effective resistance d.

Definition 2.32. For $x, y, z \in V$, let

$$M_y(x, z) := \frac{1}{2}(d(x, y) + d(y, z) - d(x, z)) \tag{2.30}$$

be the matrix of defects occurring in the triangle inequality when using the intermediate point y. Furthermore, let

$$A_y(x, z) := \begin{cases} 1 & , \ x = y = z \\ M_y(x, z) & , \ \text{otherwise} \end{cases} \quad \text{and} \quad b_y(x) = 1 - \delta_{xy}. \tag{2.31}$$

Proposition 2.33. *If (V, d) is an ERS with associated graph $G = (V, c)$ and $y \in V$, then the vector $c(\cdot, y) \in \mathbb{R}^V$ satisfies the equation*

$$A_y \cdot c(\cdot, y) = b_y.$$

Proof. First, note that the equation in the y-row simply states that $c(y, y) = 0$ which is true by definition of c. Since (V, d) is an ERS, we know that $M_y(x, \cdot) = \phi^{xy}$ is a solution of (D). Hence, for $x \neq y$, we have

$$-\frac{1}{c_y} = (\Delta \phi^{xy})(y) = \phi^{xy}(y) - \sum_{z \in V} \frac{c(y, z)}{c_y} \phi^{xy}(z).$$

Using $\phi^{xy}(y) = 0$ and the symmetry of c we get the equivalent equation

$$1 = \sum_{z \in V} c(y, z) \phi^{xy}(z) = \sum_{z \in V} M_y(x, z) c(z, y) = (A_y \cdot c(\cdot, y))(x).$$

$\square$

Proposition 2.34. *If (V, d) is an ERS and $y \in V$, we have*

$$\det A_y = \det(M_y {\upharpoonright}_{V'}) = \left(\sum_{\substack{T \text{ spanning} \\ \text{tree of } G}} \sqrt{\prod_{\substack{x, z \text{ adjacent} \\ \text{in } T}} c(x, z)} \right)^{-1} > 0.$$

Before we prove Proposition 2.34, let us recall Kirchhoff's matrix tree theorem [27]. For a modern adaptation, see [39, Theorem VI.29]. The Kirchhoff matrix of a graph $G = (V, c)$ is defined by

$$K(x, y) = \begin{cases} c_x & , x = y \\ -c(x, y) & , x \neq y \end{cases}.$$

Then, for any $y \in V$ and $V' := V \setminus \{y\}$, we have

$$\det(K {\upharpoonright}_{V'}) = \sum_{\substack{T \text{ spanning} \\ \text{tree of } G}} \sqrt{\prod_{\substack{x, z \text{ adjacent} \\ \text{in } T}} c(x, z)}.$$

We observe that $\Delta(x, y) = c_x^{-1} \cdot K(x, y)$ for all $x, y \in V$. Hence,

$$\det \Delta {\upharpoonright}_{V'} = \prod_{x \in V'} \frac{1}{c_x} \cdot \det(K {\upharpoonright}_{V'}) > 0 \tag{2.32}$$

Proof of Proposition 2.34. Fix $y \in V$ and let $V' = V \setminus \{y\}$. Since $M_y(x, z) = \phi^{xy}(z)$, we have for $x, z \in V'$,

$$(\Delta {\upharpoonright}_{V'} \cdot M_y {\upharpoonright}_{V'})(x, z) = \sum_{v \in V'} \Delta(x, v) M_y(v, z) = \sum_{v \in V'} \Delta(x, v) M_y(z, v)$$

$$= \sum_{v \in V'} \Delta(x, v) \phi^{zy}(v) = (\Delta \phi^{zy})(x) = \begin{cases} \frac{1}{c_x} & , x = z \\ 0 & , \text{otherwise} \end{cases}.$$

By (2.32) we have $\det(\Delta\!\restriction_{V'}) > 0$ and it follows that

$$\det(M_y\!\restriction_{V'}) = \frac{\det(\Delta\!\restriction_{V'} \cdot M_y\!\restriction_{V'})}{\det(\Delta\!\restriction_{V'})} = \prod_{x \neq y} \frac{1}{c_x} \cdot \det(\Delta\!\restriction_{V'})^{-1} = (\det(K\!\restriction_{V'})^{-1}.$$

The fact that $\det A_y = \det(M_y\!\restriction_{V'})$ follows from applying Laplace expansion to the y-row of A_y. $\qquad\square$

The combination of Propositions 2.33 and 2.34 gives a necessary condition for a metric space to be an ERS.

Corollary 2.35. *If a finite metric space (V, d) is an ERS, we have $\det A_y > 0$ for all $y \in V$. Furthermore, the unique solution $(c(x, y))_{x,y \in V}$ of the family of linear equation systems*

$$A_y \cdot c(\cdot, y) = b_y \ , \ y \in V \qquad\qquad \text{(LES)}$$

has non-negative entries.

Remark 2.36. Note that $\det A_y > 0$ also implies that there exists a ***unique*** solution for the linear equation system (LES). Hence, for every effective resistance on a finite set V there exists exactly **one** graph $G = (V, c)$ which induces this effective resistance.

In particular, for $x^* \in V$, the graph defined by the Star-Mesh transform (Theorem 2.6) is the only graph with vertex set $V \setminus \{x^*\}$ inducing the effective resistance $R\!\restriction_{V\setminus\{x^*\}}$. Hence, the Star-Mesh transform is the only way to remove a vertex without changing the effective resistances between all other vertices.

The uniqueness also follows from [20, Theorem 1.13]. Indeed, the theorem states that $R_G = R_H$ if and only if $\mathcal{E}_G = \mathcal{E}_H$ where $\mathcal{E}_G$ and $\mathcal{E}_H$ are energy forms associated to graphs G and H, respectively.

Remark 2.37. Our proof of Proposition 2.33 mostly uses basic graph theory and linear algebra. An alternative approach using operator theory would be to leverage the fact that $M_y : V \times V \to \mathbb{R}$ has been identified in [26, 31] as the Green function with 0-Dirichlet boundary condition on $\{y\}$.

2.7 Algebraic Characterization of Effective Resistances

In this section, we will show that the condition stated in Corollary 2.35 is not only necessary but also sufficient for a finite metric space (V, d) to an ERS. From now on we assume that $\det A_y > 0$ for all $y \in V$ and that the matrix $(c(x, y))_{x,y \in V}$ satisfying (LES) has non-negative entries.

We need to prove two things. First, that c actually defines a graph and second, that the effective resistance of this graph is given by d. For the first step, the difficult part is proving that c is symmetric. In order to do so, we will first prove that the determinant of A_y is actually independent of y.

For $x, y \in V, x \neq y$ let $T_{xy} \in \mathbb{R}^{V \times V}$ be defined by

$$T_{xy}(w, z) = \begin{cases} 1 & , w = z, w \notin \{x, y\} \\ -1 & , w \neq y, z = y \text{ or } w = y, z = x \\ 0 & , \text{ otherwise} \end{cases} \qquad (2.33)$$

Lemma 2.38. *We have* $\det T_{xy} = -1$ *and* $T_{xy}^{-1} = T_{yx}$.

Proof. To see that $\det T_{xy} = -1$, add all columns which are not indexed by x or y to the y-column and swap the x- and y-column. The resulting matrix is a diagonal matrix with entries -1 at exactly two positions and 1 otherwise. The fact that $T_{xy}^{-1} = T_{yx}$ is a simple computation. $\qquad\square$

Proposition 2.39. *For* $x, y \in V$, $x \neq y$, *we have*

$$T_{xy} A_x (T_{xy})^t = A_y \tag{2.34}$$

and

$$\det M_x = \det A_x = \det A_y = \det M_y. \tag{2.35}$$

Proof. For simplicity, let $T = T_{xy}$. Since $(A_x)^t = A_x$, it is clear that $T A_x T^t$ is also symmetric. We now compute

$$(T A_x T^t)(w, z) = \sum_{a \in V} \sum_{b \in V} T(w, a) \cdot A_x(a, b) \cdot T(z, b) =: \spadesuit$$

for all $w, z \in V$.

Case 1: $w = y = z$.

$$\spadesuit = \sum_b T(y, x) \cdot A_x(x, b) \cdot T(y, b) = (-1) \cdot 1 \cdot (-1) = 1 = A_y(y, y)$$

Case 2: $w = x = z$.

$$\spadesuit = \sum_b T(x, y) A_x(y, b) T_x y(x, b) = (-1) \cdot A_x(y, y) \cdot (-1) = d(x, y) = A_y(x, x)$$

Case 3: $w = z, w \notin \{x, y\}$:

$$\spadesuit = \sum_b T(w, w) A_x(w, b) T(w, b) + \sum_b T(w, y) A_x(y, b) T(w, b)$$
$$= A_x(w, w) - A_x(w, y) - (A_x(y, w) - A_x(y, y))$$
$$= d(x, w) - (d(x, w) + d(x, y) - d(w, y)) + d(x, y) = d(w, y) = A_y(w, w)$$

Case 4: $w = y, z \neq y$.

$$\spadesuit = \sum_b T(y, x) A_x(x, b) T(z, b) = (-1)(A_x(x, y) T(z, y) + A_x(x, z) T(z, z))$$
$$= -\underbrace{A_x(x, z)}_{=\delta_{xz}} \underbrace{T(z, z)}_{=1-\delta_{xz}} = 0 = A_y(y, z)$$

Case 5: $w = x, z \notin \{x, y\}$.

$$\spadesuit = \sum_b T(x, y) A_x(y, b) T(z, b) = -(A_x(y, z) - A_x(y, y))$$
$$= d(x, y) - \frac{1}{2}(d(x, y) + d(x, z) - d(y, z)) = \frac{1}{2}(d(x, y) + d(y, z) - d(x, z))$$
$$= A_y(x, z)$$

Case 6: $w \neq z, w, z \notin \{x, y\}$.

$$\spadesuit = \sum_b A_x(w, b)T(z, b) - \sum_b A_x(y, b)T(z, b)$$
$$= (A_x(w, z) - A_x(w, y)) - (A_x(y, z) - A_x(y, y))$$
$$= \frac{1}{2}\left[d(x, z) - d(x, y) - d(w, z) + d(w, y) - d(x, y) - d(x, z) + d(y, z)\right] + d(x, y)$$
$$= \frac{1}{2}(d(w, y) + d(y, z) - d(w, z)) = A_y(w, z)$$

This concludes the proof of (2.34). (2.35) now follows using Laplace expansion to obtain $\det A_y = \det M_y$ and Lemma 2.38 to see

$$\det A_y = \det(T_{xy}A_x(T_{xy})^t) = \det T_{xy} \det A_x \det T_{xy} = (-1)\det A_x(-1) = \det A_x.$$

$\square$

In what follows, we denote by A_{xy} be the matrix that results from replacing the y-column in A_x with b_x. Then, Cramer's rule [7, Section 5.13] states that

$$c(y, x) = \frac{\det A_{xy}}{\det A_x}.$$

From Proposition 2.39 we already know how the transformation $A \mapsto T_{xy}A(T_{xy})^t$ acts on the matrix A_x. Now we will investigate its effects on A_{xy}.

Lemma 2.40. *Let $x, y \in V$, $x \neq y$ and $A := T_{xy}A_{xy}T_{xy}^t$. Then,*

$$A(w, x) = \delta_{wx} \text{ and } A(w, y) = \delta_{wy}$$

holds for all $w \in V$.

Proof. For simplicity let $T = T_{xy}$. For $z = x$ or $z = y$ and $b \in V$, we have

$$T(z, b) = \begin{cases} -1 & , z = x, b = y \text{ or } z = y, b = x \\ 0 & , \text{ otherwise} \end{cases}$$

We consider three cases and compute

$$A(w, z) = \sum_{a \in V} \sum_{b \in V} T(w, a)A_{xy}(a, b)T(z, b)$$

Case 1: $w = x$.

$$A(x, x) = A_{xy}(y, y) = b_x(y) = 1$$
$$A(x, y) = A_{xy}(y, x) = 0$$

Case 2: $w = y$.

$$A(y, x) = A_{xy}(x, y) = b_x(x) = 0$$
$$A(y, y) = A_{xy}(x, x) = 1$$

Case 3: $w \notin \{x, y\}$.

$$A(w, x) = -A_{xy}(w, y) + A_{xy}(y, y) = b_x(y) - b_x(w) = 0$$
$$A(w, y) = -A_{xy}(w, x) + A_{xy}(y, x) = 0$$

$\square$

Proposition 2.41. *Let* $x, y \in V$, $x \neq y$, $A := T_{xy}A_{xy}T_{xy}^t$ *and* $B := T_{yx}A_{yx}T_{yx}^t$. *Then* $A(w, z) = B(z, w)$ *for all* $w, z \notin \{x, y\}$ *and it follows that*

$$\det A = \det B.$$

Proof. We have

$$
\begin{aligned}
A(w, z) &= \sum_{a \in V} \sum_{b \in V} T_{xy}(w, a) A_{xy}(a, b) T_{xy}(z, b) \\
&= \sum_{b} A_{xy}(w, b) T(z, b) - A_{xy}(y, b) T(z, b) \\
&= A_{xy}(w, z) - \underbrace{A_{xy}(w, y)}_{=1} - (A_{xy}(y, z) - \underbrace{A_{xy}(y, y)}_{=1}) \\
&= \frac{1}{2}(d(x, w) + d(y, z) - d(w, z) - d(y, x)) \\
&= A_{yx}(z, w) - A_{yx}(x, w) - (\underbrace{A_{yx}(z, x)}_{=1} - \underbrace{A_{yx}(x, x)}_{=1}) \\
&= \sum_{b} A_{yx}(z, b) T_{yx}(w, b) - A_{yx}(x, b) T_{yx}(w, b) \\
&= \sum_{a} \sum_{b} T_{yx}(z, a) A_{yx}(a, b) T_{yx}(w, b) = B(z, w)
\end{aligned}
$$

Using Lemma 2.40 and Laplace expansion we have

$$\det A = 1 \cdot 1 \cdot \det\left((A(w, z))_{w,z \in V \setminus \{x,y\}}\right) = 1 \cdot 1 \cdot \det\left(B(z, w)_{z,w \in V \setminus \{x,y\}}\right) = \det B$$

$\square$

We can now combine these computations to obtain the symmetry of the solution matrix c of (LES).

Proposition 2.42. *For* $x, y \in V$ *we have* $c(x, y) = c(y, x)$.

Proof. For $x = y$ there is nothing to show. For $x \neq y$, it follows from Proposition 2.41 that

$$\det A_{xy} = \det(T_{xy}A_{xy}T_{xy}^t) = \det(T_{yx}A_{yx}T_{yx}^t) = \det A_{yx}.$$

Hence,

$$c(x, y) = \frac{\det A_{yx}}{\det A_y} = \frac{\det A_{xy}}{\det A_x} = c(y, x).$$

$\square$

Since c is a symmetric matrix with a vanishing diagonal and non-negative entries, it defines a graph $G = (V, c)$. At this point we do not know whether this graph is connected. We do know, however, that it has no isolated vertices since this would correspond to a column in c consisting only of zeros. This is clearly impossible due to the form of (LES). Since there are no isolated vertices, the Laplacian of G is well-defined. By Lemma 2.2, (D) has a solution if and only if G is connected and we will show that $\phi^{xy}(z) := M_y(x, z)$ is such a solution for the Laplacian of G, implying that G is connected.

Proposition 2.43. *For $x, y \in V$, $x \neq y$, let $\phi^{xy}(z) := M_y(x, z)$. Then,*

$$\Delta \phi^{xy} = \frac{1}{c_x} \mathbf{1}_x - \frac{1}{c_y} \mathbf{1}_y$$

$$\phi^{xy}(y) = 0 .$$

In particular, $d(x, y)$ is the effective resistance between x and y in the graph (V, c).

In order to prove this proposition, we need to observe some basic properties of M_y. For $w, x, z \in V$, we have

$$M_y(x, x) = d(x, y) = M_x(y, y) \tag{2.36}$$

$$M_y(x, z) = M_y(z, x) \tag{2.37}$$

$$M_y(x, z) = d(x, y) - M_x(y, z) \tag{2.38}$$

$$M_y(x, w) - M_y(x, z) = M_w(y, z) - M_w(x, z) \tag{2.39}$$

$$= M_z(x, w) - M_z(y, w)$$

Proof of Proposition 2.43. By (LES), we have $c(x, x) = 0$ and

$$1 = (A_y \cdot c(\cdot, y))(x) = \sum_{w \in V} M_y(x, w) c(w, y)$$

whenever $x \neq y$. Using (2.36) and (2.38), we compute

$$c_x \cdot \Delta \phi^{xy}(x) = c_x \phi^{xy}(x) - \sum_w c(x, w) \phi^{xy}(w) = c_x M_y(x, x) - \sum_w c(w, x) M_y(x, w)$$

$$= c_x d(x, y) - \sum_w c(w, x)(d(x, y) - M_x(y, w)) = \sum_w M_x(y, w) c(w, x)$$

$$= 1.$$

For $z \notin \{x, y\}$, we have

$$1 = \sum_w M_z(x, w) c(w, z) = \sum_w M_z(y, w) c(w, z).$$

Combining this with (2.39), we get

$$c_z(\Delta \phi^{xy})(z) = c_z \phi^{xy}(z) - \sum_w c(z, w) \phi^{xy}(w) = c_z M_y(x, z) - \sum_w c(w, z) M_y(x, w)$$

$$= \sum_w c(w, z)[M_y(x, z) - M_y(x, w)] = \sum_w c(w, z)[M_z(y, w) - M_z(x, w)]$$

$$= 1 - 1 = 0.$$

Lastly, $M_y(x, y) = 0$ yields

$$c_y(\Delta\phi^{xy})(y) = c_y\phi^{xy}(y) - \sum_w c(y, w)\phi^{xy}(w) = c_y M_y(x, y) - \sum_w c(w, y)M_y(x, w)$$

$$= 0 - 1 = -1.$$

Now that we have established that $\phi^{xy} = M_y(x, \cdot)$ solves the Dirichlet problem (D) on G, it is a direct consequence that the effective resistance R of G is in fact the metric d.

$$R(x, y) = \phi^{xy}(x) = M_y(x, x) = d(x, y).$$

$\square$

Combining all results of this section, we obtain a concise characterization of finite effective resistance spaces.

Theorem 2.44. *Let (V, d) be a finite metric space and let M_y, A_y, b_y be as in (2.30) and (2.31). Then, d is the effective resistance of a graph $G = (V, c)$ if and only if*

$$\det M_y > 0$$

for some $y \in V$ and the family of linear equation systems

$$A_y \cdot c(\cdot, y) = b_y \ , \ y \in V$$

admits a solution matrix $(c(x, y))_{x,y\in V}$ with non-negative entries.

Example 2.45. We give an example for a naturally occurring metric space which is not representable as an effective resistance. Let $V = \{v_0, \ldots, v_3\}$ and consider the graph $C_4 = (V, c)$ where

$$c(v_i, v_j) = \begin{cases} 1 & , |i - j| \in \{1, 3\} \\ 0 & , \text{ otherwise} \end{cases}.$$

The geodesic metric of C_4 is given by

$$(d_{C_4}(v_i, v_j))_{i,j\in\{0,\ldots,3\}} = \begin{pmatrix} 0 & 1 & 2 & 1 \\ 1 & 0 & 1 & 2 \\ 2 & 1 & 0 & 1 \\ 1 & 2 & 1 & 0 \end{pmatrix}$$

Hence,

$$M_{v_0} = \begin{pmatrix} 1 & 1 & 0 \\ 1 & 2 & 1 \\ 0 & 1 & 1 \end{pmatrix}$$

and $\det M_{v_0} = 0$. Hence, (V, d_{C_4}) is no effective resistance space.

Example 2.46. This example demonstrates that the second condition in Theorem 2.44, namely that the family of linear equation systems admits a solution with non-negative entries, can not be dropped. Moreover, it does not suffice to check $c(\cdot, y) \geq 0$ for only one $y \in V$. Consider the metric space $(\{v_0, v_1, v_2, v_3\}, d)$ where d is given by

$$(d(v_i, v_j))_{i,j \in \{0,\dots,3\}} = \begin{pmatrix} 0 & 7/2 & 2 & 1 \\ 7/2 & 0 & 1 & 2 \\ 2 & 1 & 0 & 1 \\ 1 & 2 & 1 & 0 \end{pmatrix}.$$

It follows that

$$M_{v_0} = \frac{1}{4} \begin{pmatrix} 14 & 9 & 5 \\ 9 & 8 & 4 \\ 5 & 4 & 4 \end{pmatrix}$$

and thus $\det M_{v_0} = \frac{15}{16} > 0$. Hence, the family of linear equation systems (LES) has a unique solution matrix

$$(c(v_i, v_j))_{i,j \in \{0,\dots,3\}} = \frac{1}{15} \begin{pmatrix} 0 & -4 & 4 & 16 \\ -4 & 0 & 16 & 4 \\ 4 & 16 & 0 & 11 \\ 16 & 4 & 11 & 0 \end{pmatrix}$$

and we see that $c(v_0, v_1) < 0$. This is also the only negative entry of c (apart from $c(v_1, v_0)$).

2.8 Graph Realizations of Metrics

2.8.1 Geodesic Graphs of Finite Metric Spaces

We review a method of realizing finite metric spaces using graphs introduced in 1965 by Hakimi and Yau [17]. For the convenience of the reader, we reproduce some of the results and sketch the proofs.

We say a graph $G = (V, c)$ **realizes a metric space** (X, d) if $X = V$ and the d is the geodesic metric of G. Note that in [17] a realization is allowed to have more vertices than the metric space, i.e. Hakimi and Yau only demand $X \subseteq V$. In our work, we only consider realizations which satisfy $X = V$. In what follows, let (V, d) be a finite metric space.

Our first observation is that every finite metric space can be realized using a complete graph.

Lemma 2.47. *The metric space (V, d) is realized by the graph $G = (V, c)$ with*

$$c(x, y) = \frac{1}{d(x, y)} \quad \forall x \neq y.$$

There may exist multiple graphs realizing the same metric space, cf. Example 2.28. This is due to **redundant edges**, i.e. edges which can be removed from a realization without changing its geodesic metric. A realization which has no redundant edges is called *irreducible* [17]. Hakimi and Yau showed that for any finite metric space, there exists a unique irreducible realization. Later, Goldman [15] proved that a realization (V, c) of a metric space (V, d) is irreducible if and only if

$$c(x, y) = \begin{cases} 1/d(x,y) & , d(x, y) < \min_{z \in V \setminus \{x,y\}} (d(x, z) + d(z, y)) \\ 0 & , \text{otherwise} \end{cases} . \qquad (2.40)$$

For $x, y \in V$, $x \neq y$, we define the set of *intermediate points* $T(x, y)$ between x and y by

$$T(x, y) := \{z \in V \setminus \{x, y\} \mid d(x, y) = d(x, z) + d(z, y)\} .$$

Using the notion of intermediate points, we can write (2.40) in the following equivalent form.

$$c(x, y) = \begin{cases} 1/d(x,y) & , T(x, y) = \emptyset \\ 0 & , T(x, y) \neq \emptyset. \end{cases} \qquad (2.41)$$

Definition 2.48. For a finite metric space (V, d), let $G_d := (V, c_d)$ where c_d is defined as in (2.41). We call G_d the *geodesic graph* of (V, d).

Note that for any $x, y \in V$ and any shortest path $\gamma : x \to y$ in G_d, we have $z \in T(x, y)$ for all $z \in \gamma \setminus \{x, y\}$.

Example 2.49. Consider $\mathbb{Z}^2$ with the metric induced by the l_1 norm, i.e.

$$d((x_1, y_1), (x_2, y_2)) = |x_1 - x_2| + |y_1 - y_2|$$

and let $V = \{(x, y) \in \mathbb{Z}^2 \mid |x| + |y| \leq 2\}$. The geodesic graph for (V, d) is shown in Figure 2.7.

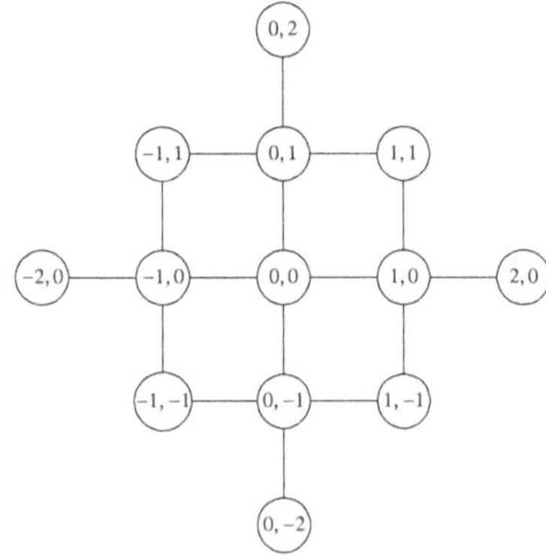

Figure 2.7: The geodesic graph of a subset of $\mathbb{Z}^2$

It is important to note that an arbitrary graph is in general not the geodesic graph for its own geodesic metric, see Figure 2.8. However, every tree has this property.

Figure 2.8: A graph G (left) and the geodesic graph for d_G (right).

Lemma 2.50. *If G is a tree, then G is the geodesic graph for (V, d_G). In particular, G_{d_G} is a tree.*

Proof. By definition, $G = (V, c)$ realizes (V, d_G). Since G contains no cycles, there exists exactly one simple path $x \to y$ for any two vertices $x, y \in G$. Hence, we have $z \in T(x, y)$ if and only if this path uses z. It follows that $c(x, y) = 0$ if and only if $T(x, y) \neq \emptyset$ which is equivalent to $c_{d_G}(x, y) = 0$. If $c(x, y) > 0$, we have

$$\frac{1}{c(x, y)} = \inf_{\gamma \in \Gamma_G(x, y)} r(\gamma) = d_G(x, y) = \frac{1}{c_{d_G}(x, y)}.$$

Hence, $G = G_{d_G}$. $\square$

2.8.2 Geodesic Graphs of Effective Resistance Spaces

Since geodesic graphs are unique graph realization of their respective metric spaces, they can act as a method of visualizing finite metric spaces. Now we will investigate what a metric space has to "look like" in order to be an effective resistance space (ERS).

Let G be a graph with effective resistance R. Due to the Star-Mesh transform, every metric subspace of (V, R) is again an ERS (Proposition 2.26). Furthermore, by Proposition 2.27, we know that the triangle inequality in (V, R) is an equality if and only if the intermediate point is a **cut-point** in G. This means that removing this vertex (and all incident edges) from G results in an unconnected graph. This will yield a first necessary condition for realizations of ERS.

Example 2.51. Consider the geodesic metric of a cycle graph on four vertices as in Example 2.45, namely $(\{v_0, v_1, v_2, v_3\}, d)$ where

$$(d(v_i, v_j))_{i,j \in \{0,\ldots,3\}} = \begin{pmatrix} 0 & 1 & 2 & 1 \\ 1 & 0 & 1 & 2 \\ 2 & 1 & 0 & 1 \\ 1 & 2 & 1 & 0 \end{pmatrix}.$$

Suppose there is some graph $G = (V, c)$ such that $R_G = d$. Note that

$$d(v_0, x) + d(x, v_2) - d(v_0, v_2) = 0 \text{ and}$$
$$d(v_1, y) + d(y, v_3) - d(v_1, v_3) = 0$$

for $x \in \{v_1, v_3\}$ and $y \in \{v_0, v_2\}$. By Proposition 2.27, this implies that all paths $v_0 \to v_2$ in G use v_1 as well as v_3 and all paths $v_1 \to v_3$ in G use v_0 as well as v_2. This is impossible and it follows that there exists no graph G such that $R_G = d$.

Lemma 2.52. *Let $G = (V, c)$ be a weighted graph with effective resistance R. Furthermore, let $G_R = (V, c_R)$ be the geodesic graph of (V, R). For $x, y \in V$, we then have*

$$c_R(x, y) = 0 \Rightarrow c(x, y) = 0,$$

i.e. if there exists no edge between x and y in the geodesic graph G_R, then there exists no edge between x and y in G.

Proof. Let $x, y \in V$, $x \neq y$, such that $c_R(x, y) = 0$. Hence, there exists $z \in T(x, y)$, i.e.

$$R(x, z) + R(z, y) = R(x, y).$$

By Proposition 2.27, it follows that all paths $x \to y$ in G use z. In particular, $c(x, y) = 0$. $\square$

Example 2.53. Consider a metric space (V, d), $V = \{v_1, \dots, v_4\}$, with geodesic graph $G_d = (V, c_d)$ and let $c_{ij} := c_d(v_i, v_j)$ for convenience. Assume that $(v_1, v_2), (v_2, v_3), (v_3, v_4), (v_4, v_1) \in E(G_d)$, $c_{13} = 0$ and $c_{24} \geq 0$, see Figure 2.9.

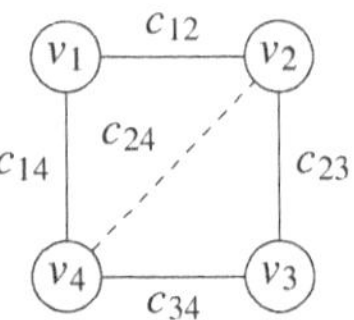

Figure 2.9: The geodesic graph G_d for (V, d) (Example 2.53).

We claim that (V, d) is no ERS. Indeed, assume that there is a graph $G = (V, c)$ such that $R_G = d$. Since $c_{13} = 0$, it follows that v_2 or v_4 are in $T(v_1, v_3)$. W.l.o.g. let

$$d(v_1, v_2) + d(v_2, v_3) - d(v_1, v_3) = 0.$$

By Proposition 2.27, all paths $v_1 \to v_3$ in G use v_2. Hence, G can only be of the shapes shown in Figure 2.10. Shapes 3 to 7 are trees and by Lemma 2.50, this implies that G_d is

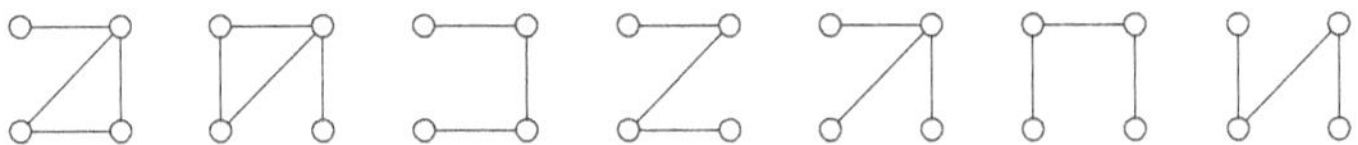

Figure 2.10: Possible shapes of G (Example 2.53).

also a tree which is not the case. If G has shape 1, all paths $v_1 \to v_4$ use v_2. Hence,

$$d(v_1, v_2) + d(v_2, v_4) - d(v_1, v_4) = 0$$

and it follows that $c_{14} = 0$ which is a contradiction. The analogous argument holds for shape 2. This proves the claim.

2.8.3　Incomplete Cycles

In this section we will investigate the connection between the geodesic graph of a metric space and the geodesic graphs of its metric subspaces.

Lemma 2.54. *Let $W \subseteq V$ and $d' := d{\restriction}_{W \times W}$. Furthermore let $G_d = (V, c_d)$ and $G'_d = (W, c'_d)$ be the geodesic graphs for (V, d) and (W, d'), respectively. Then,*

$$\forall\, x, y \in W : c_d(x, y) > 0 \Rightarrow c'_d(x, y) > 0$$

or equivalently

$$\forall\, x, y \subset W : c'_d(x, y) = 0 \Rightarrow c_d(x, y) = 0.$$

Proof. Follows immediately from Definition 2.48.　　　　　　　　　　　　　　　□

Definition 2.55. Let $G = (V, c)$ be any graph, $n \in \mathbb{N}$ and $\mathfrak{c} = (x_0, \dots, x_n)$ a proper cycle in G.

1. We say $\mathfrak{c}$ is an ***incomplete (n-)cycle*** iff there exists $0 \le i < j \le n - 1$ such that $c(x_i, x_j) = 0$.

2. We say $\mathfrak{c}$ is a ***shortest incomplete cycle*** iff $\mathfrak{c}$ is an incomplete cycle whose sum of edge-wise resistances is minimal among incomplete cycles, i.e.

$$r(\mathfrak{c}) = \min \left\{ r(\mathfrak{c}') \mid \mathfrak{c}' \text{ incomplete cycle in } G \right\}.$$

　　There are a few things to note regarding the definition of incomplete cycles. First, since $\mathfrak{c}$ is a proper cycle, we have $x_i \ne x_j$ for all $0 \le i < j \le n - 1$. Secondly, there exist no incomplete n-cycles for $n = 3$. Thirdly, being an incomplete cycle means that the subgraph of G induced by the vertices of $\mathfrak{c}$ is not a complete graph. Lastly, note that there may exist incomplete cycles which visit fewer vertices than a shortest incomplete cycle and that the there may exists several shortest incomplete cycles.

　　We say a cycle $(x_0, \dots, x_n)$ contains an ***interior edge*** if there exist i, j such that $2 \le |i - j| \le n - 2$ and $(x_i, x_j) \in E(G)$.

Lemma 2.56. *Let (V, d) be a finite metric space and $\mathfrak{c} = (x_0, \dots, x_n)$ a shortest incomplete cycle in the geodesic graph G_d. If $\mathfrak{c}$ contains an interior edge, then $\mathfrak{c}$ is a 4-cycle.*

Proof. Let $i < j$ be such that $c(x_i, x_j) > 0$ and $2 \le |i - j| < n - 2$, i.e. there is an interior edge in $\mathfrak{c}$ between x_i and x_j. Then $\mathfrak{c}_1 := (x_0, \dots, x_i, x_j, \dots, x_n)$ and $\mathfrak{c}_2 := (x_i, \dots, x_j, x_i)$ are two cycles in G_d such that $r(\mathfrak{c}_i) < r(\mathfrak{c})$, $i = 1, 2$, by definition of G_d.

　　Since $r(\mathfrak{c})$ is minimal among all incomplete cycles, it follows that $\mathfrak{c}_1$ and $\mathfrak{c}_2$ both induce complete subgraphs. By incompleteness of $\mathfrak{c}$, there exist x_k in $\mathfrak{c}_1$ and x_l in $\mathfrak{c}_2$ such that $c(x_k, x_l) = 0$. Thus, $\widetilde{\mathfrak{c}} := (x_0, \dots, x_i, x_l, x_j, \dots, x_n)$ is an incomplete cycle in G_d with

$$r(\mathfrak{c}) \le r(\widetilde{\mathfrak{c}}) \le r(\mathfrak{c}).$$

Hence, $\widetilde{\mathfrak{c}} = \mathfrak{c}$ and it follows that $\mathfrak{c}_2 = (x_i, x_l, x_j, x_i)$ is a 3-cycle. By the same argument, $\mathfrak{c}_1$ is a 3-cycle and it follows that $\mathfrak{c} = (x_k, x_i, x_l, x_j)$ is a 4-cycle.　　　　　　□

Lemma 2.57. *Let $c = (x_0, \ldots, x_n)$ be a shortest incomplete cycle in a geodesic graph G_d. Then, for all $0 \le i < j \le n - 1$ such that $c_d(x_i, x_j) = 0$, some shortest path $x_i \to x_j$ is part of c. In particular, $x_l \in T(x_i, x_j)$ for all $i < l < j$ or for all $0 \le l < i$ and $j < l \le n$.*

Proof. The main idea is the following. If a path γ which is not part of the cycle c had a smaller r-length than either part of c, we could substitute one part of c for γ and obtain an incomplete cycle (since $c_d(x_i, x_j) = 0$) which would be strictly shorter than c. However, this would be a contradiction to c being a shortest incomplete cycle. In the following, we give a more detailed version of this proof.

Let $c = (x_0, \ldots, x_n)$, $i < j$ such that $c_d(x_i, x_j) = 0$ and γ be a shortest path $x_j \to x_i$ in G_d, i.e. $r(\gamma) = d(x_i, x_j)$. Furthermore, let $c_1 = (x_i, \ldots, x_j)$ and $c_2 = (x_j, \ldots, x_n, x_1, \ldots, x_i)$ be the two parts of c connecting x_i and x_j. Note that γ does not visit any vertex twice because it is a shortest path and neither do c_1, c_2 since $c = (c_1, c_2)$ is a proper cycle.

If $r(\gamma) = r(c_2)$, then c_2 is a shortest path $x_j \to x_i$ and we are done.

From now on, let $r(\gamma) < r(c_2)$ and $c' := (c_1, \gamma)$. Then, c' is a cycle with $r(c') < r(c)$. Furthermore, $c_d(x_i, x_j) = 0$ implies $L(c') = L(c_1) + L(\gamma) \ge 4$ and that the subgraph induced by c' is not complete. Since c is a shortest incomplete cycle, c' can not be proper, i.e. there exists at least one vertex v distinct from x_i, x_j which is visited both by c_1 and γ.

If $n = 4$, we have $c_1 = (x_i, v, x_j)$. At v, we split γ into two parts $\gamma_1 : x_i \to v$ and $\gamma_2 : v \to x_j$ and get

$$d(x_i, x_j) \le r(c_1) = d(x_i, v) + d(v, x_j) \le r(\gamma_1) + r(\gamma_2) = r(\gamma) = d(x_i, x_j)$$

by definition of G_d. Hence, c_1 is a shortest path $x_i \to x_j$.

For $n \ge 5$, it follows from Lemma 2.56 that c has no interior edges. Let $c_1 \cap \gamma$ be the set of intersection points of c_1 and γ. If all vertices of c_1 are in $c_1 \cap \gamma$, the definition of G_d implies that $r(c_1) \le r(\gamma)$ and it follows that c_1 is a shortest path $x_i \to x_j$.

Now consider the case where at least one vertex x_k of c_1 is not visited by γ. At the vertices in $c_1 \cap \gamma$, we can split c' into a finite number of smaller cycles which are proper or of length 2. Let $\widetilde{c}$ denote the proper cycle containing x_k. Then, $\widetilde{c}$ must contain the path (x_{k-1}, x_k, x_{k+1}) and since $\widetilde{c}$ consists of subpaths of c', we have $r(\widetilde{c}) \le r(c') < r(c)$. By minimality of c among incomplete cycles, it follows that $\widetilde{c}$ must induce a complete subgraph. Hence, $c_d(x_{k-1}, x_{k+1}) > 0$ which is a contradiction to c not containing interior edges.

$\square$

Proposition 2.58. *Let $G_d = (V, c)$ be the geodesic graph of (V, d). If G_d contains any incomplete n-cycles, then there exists $W \subseteq V$, $|W| = 4$, such that the geodesic graph $G'_d = (W, c')$ of $(W, d\restriction_{W \times W})$ is an incomplete 4-cycle.*

Proof. Let $c = (x_0, \ldots, x_n)$ be a shortest incomplete cycle in G_d. First, suppose that c is a 4-cycle. In this case, let $W = \{x_0, \ldots, x_3\}$. By Lemma 2.54, we know that G'_d contains all edges that are contained in c, i.e. it is again a cycle. W.l.o.g. assume that $c(x_0, x_2) = 0$. By Lemma 2.57, we have $x_1 \in T(x_0, x_2)$ or $x_3 \in T(x_0, x_2)$. Hence, $c'(x_0, x_2) = 0$ making $G_{d'}$ an incomplete cycle.

Now suppose that $n \geq 5$. By Lemma 2.56, there exist no interior edges in c. Before we choose $0 \leq i_0 < i_1 < i_2 < i_3 \leq n - 1$ such that $W = \{x_{i_0}, x_{i_1}, x_{i_2}, x_{i_3}\}$, we note that every such choice of vertices will yield

$$c'(x_{i_0}, x_{i_2}) = c'(x_{i_1}, x_{i_3}) = 0$$

since $\{x_{i_1}, x_{i_3}\} \cap T(x_{i_0}, x_{i_2}) \neq \emptyset$ and $\{x_{i_0}, x_{i_2}\} \cap T(x_{i_1}, x_{i_3}) \neq \emptyset$ by Lemma 2.57. It follows that for every choice of i_j, G'_d will not be a complete graph. Since G'_d is connected, there will be at least three edges in it.

All that is left to show is that there exists a choice of $i_0, \ldots, i_3$ such that all four 'outer' edges exist. In order to do so, we choose $i_0, \ldots, i_3$ such that the parts of c connecting x_{i_j} and $x_{i_{j+1}}$ (where $i_4 := i_0$) are shortest paths. Note that by definition of G_d, we have

$$2r(x_i, x_{i+1}) = 2d(x_i, x_{i+1}) < d(x_i, x_{i+1}) + d(x_{i+1}, x_{i+2}) + d(x_{i+2}, x_i)$$
$$\leq \ldots \leq d(x_i, x_{i+1}) + r(x_{i+1}, \ldots, x_n) + r(x_0, \ldots, x_i) = r(c).$$

Hence, $r(x_i, x_{i+1}) < \frac{1}{2}r(c)$ for all $i = 0, \ldots, n - 1$, which implies that we can choose $i_0, \ldots, i_3$ such that

$$r(x_{i_j}, \ldots, x_{i_{j+1}}) < \frac{1}{2}r(c), \ j = 0, 1, 2, 3,$$

where $r(x_{i_3}, \ldots, x_{i_4}) := r(x_{i_3}, \ldots, x_n) + r(x_0, \ldots, x_{i_0})$. It follows that

$$r(x_{i_j}, \ldots, x_{i_{j+1}}) < \frac{1}{2}r(c) < r(c) - r(x_{i_j}, \ldots, x_{i_{j+1}})$$

and therefore $r(x_{i_j}, \ldots, x_{i_{j+1}}) = d(x_{i_j}, x_{i_{j+1}})$ by Lemma 2.57.

Now suppose we know that the outer edges between x_{i_1}, x_{i_2} and x_{i_3} exist in G'_d. The only way that $c'(x_{i_0}, x_{i_1}) = 0$ is that

$$r(x_{i_0}, \ldots, x_{i_1}) = d(x_{i_0}, x_{i_1}) = d(x_{i_1}, x_{i_2}) + d(x_{i_2}, x_{i_3}) + d(x_{i_3}, x_{i_0})$$
$$= r(x_{i_1}, \ldots, x_{i_2}) + r(x_{i_2}, \ldots, x_{i_3}) + r(x_{i_3}, \ldots, x_{i_0})$$

which is a contradiction to the inequality above. Hence, $c'(x_{i_0}, x_{i_1}) > 0$ making G'_d an incomplete 4-cycle.

$\square$

Theorem 2.59. *Let (V, d) be an ERS. Then, G_d contains no incomplete cycles.*

Proof. Assume that there exists an incomplete cycle in G_d. By Proposition 2.58, there exists $W \subseteq V$, $|W| = 4$, such that the geodesic graph of $(W, d\restriction_{W \times W})$ is an incomplete 4-cycle, i.e. it is isomorphic to one of the graphs shown in Figure 2.11. By Proposition

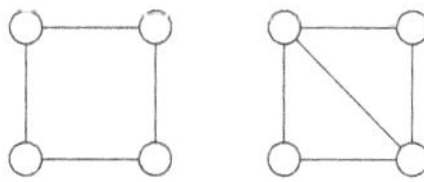

Figure 2.11: Possible shapes of an incomplete 4-cycle.

2.26, $(W, d\restriction_{W \times W})$ is an ERS. However, as shown in Example 2.53, there exists no ERS such that its geodesic graph is isomorphic to one of the graphs shown in Figure 2.11. $\square$

Theorem 2.59 gives a necessary condition for any finite metric space to be an ERS. However, it is not sufficient as the following example shows.

Example 2.60. Let $V = \{v_1, \ldots, v_4\}$ and consider a metric space (V, d) with a complete geodesic graph G_d shown in Figure 2.12 where $\lambda := d(v_1, v_3) = d(v_2, v_4)$. Note that the

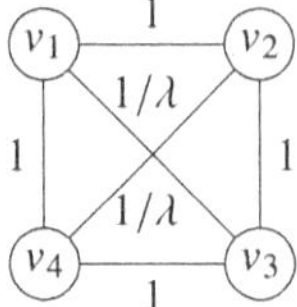

Figure 2.12: A geodesic graph

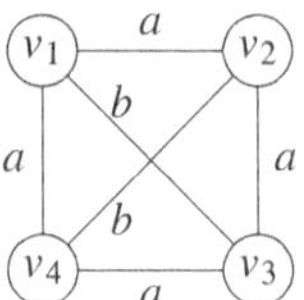

Figure 2.13: Possible graph with $R_G = d$.

completeness of G_d implies $0 < \lambda < 2$. Now assume that there exists a graph $G = (V, c)$ such that $R_G = d$. Since G is uniquely determined by d, the symmetry of G_d implies that G must be of the form shown in Figure 2.13 with $a, b \in \mathbb{R}$. Solving the linear equation systems which arise from the definition of the effective resistance (D) yields

$$a = \frac{1}{2} \cdot \frac{1}{2 - \lambda}$$
$$b = \frac{3}{2} \cdot \frac{\lambda - 4/3}{\lambda(2 - \lambda)} .$$

It follows that for any $\lambda \in (4/3, 2)$, the edge weight b is negative. Hence, for e.g. $\lambda = 1.4$, (V, d) is no ERS although its geodesic graph is complete and therefore satisfies the condition of Theorem 2.59.

2.8.4 Trees of Cliques

Definition 2.61. Let $G = (V_G, c_G)$ and $H = (V_H, c_H)$ be two graphs, $v \in V_G$ and $w \in V_H$. We define $(G, v) \oplus (H, w) = (V, c)$ to be the graph with $V = (V_G \,\dot{\cup}\, V_H) \setminus \{w\}$ and

$$c(x, y) := \begin{cases} c_G(x, y) & , x, y \in V_G \\ c_H(x, y) & , x, y \in V_H \\ c_H(w, y) & , x = v, y \in V_H \\ c_H(x, w) & , x \in V_H, y = v \\ 0 & , \text{otherwise} \end{cases} .$$

For two classes of graphs $\mathcal{A}, \mathcal{B}$ let

$$\mathcal{A} \oplus \mathcal{B} := \{(G, v) \oplus (H, w) \mid G \in \mathcal{A}, v \in V_G, H \in \mathcal{B}, w \in V_H\} .$$

Intuitively, $(G, v) \oplus (H, w)$ is the graph resulting from gluing G and H together by identifying v with w. Note that we can canonically interpret G and H as subgraphs of $(G, v) \oplus (H, w)$. Furthermore, every path in G which uses vertices in $V_G \setminus \{v\}$ as well as vertices in $V_H \setminus \{w\}$ must visit v.

Example 2.62. Figures 2.14 and 2.15 show two examples of gluing together graphs.

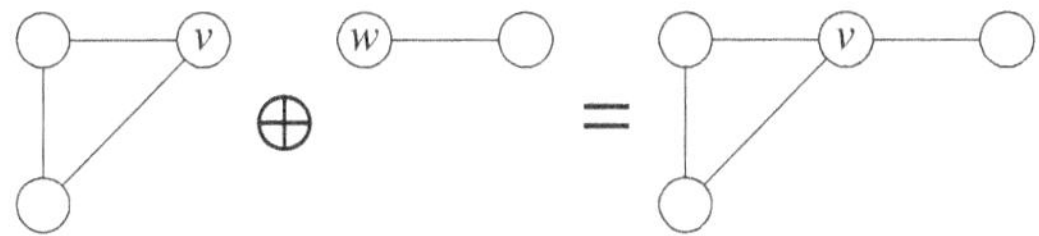

Figure 2.14: An example of gluing together two graphs.

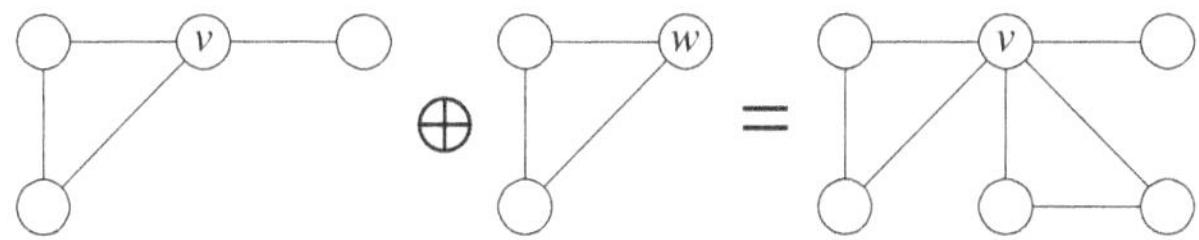

Figure 2.15: Another example of gluing together two graphs.

Definition 2.63. Let $\mathcal{G}_0$ be the class of complete graphs on a finite set of vertices. For $n > 0$ let

$$\mathcal{G}_{n+1} = \mathcal{G}_n \oplus \mathcal{G}_0$$

and

$$\mathcal{G} := \bigcup_{n \geq 0} \mathcal{G}_n.$$

We call $\mathcal{G}$ the class of **trees of cliques**.

Remark 2.64. Suppose we exchanged $\mathcal{G}_0$ in the above construction with the class of all single-edge graphs (i.e. connected graphs on two vertices). Then, we would start with a single edge and successively attach new edges. We would never create proper cycles since every new edge is attached at exactly one already present vertex. Thus, the resulting class $\mathcal{G}$ would be that of all finite trees.

Since we can attach complete graphs at every step, the resulting graph can be thought of as a tree of complete graphs, or a tree of cliques.

By definition, it is clear that $\mathcal{G}$ is closed under $\oplus$. More precisely, for any $G = (V_G, c_G), H = (V_H, c_H) \in \mathcal{G}$ and $v \in V_G, w \in V_H$, we have $(G, v) \oplus (V, w) \in \mathcal{G}$.

Lemma 2.65. *If $G \in \mathcal{G}$, then G does not contain any incomplete cycle.*

Proof. Since $G \in \mathcal{G}$ implies $G \in \mathcal{G}_n$ for some $n \geq 0$, we will use induction. For $G \in \mathcal{G}_0$, the statement is obvious. Now let $G \in \mathcal{G}_n$ for $n > 0$. Then, there exist graphs $G_1 = (V_1, c_1) \in \mathcal{G}_{n-1}$ and $G_2 = (V_2, c_2) \in \mathcal{G}_0$ and vertices $v_1 \in V_1, v_2 \in V_2$ such that

$$G = (G_1, v_1) \oplus (G_2, v_2).$$

Let $\mathfrak{c}$ be any cycle in G. Since every path in G connecting G_1 and $G_2 \setminus \{v_2\}$ must use v_1, $\mathfrak{c}$ must be completely contained in either G_1 or G_2. By induction, G_1 and G_2 do not contain incomplete cycles. Hence, $\mathfrak{c}$ cannot be incomplete. $\qquad\square$

Definition 2.66. Let $G = (V, c)$ be a graph and $W \subseteq V$. Then, W is called a ***clique*** of G, if $G \restriction_W$ is complete, i.e. if

$$c(x, y) > 0 \ \forall\, x, y \in W.$$

A clique is called ***maximal*** if there exists no bigger clique containing it.

For $W \subseteq V$, we say that a path $(x_0, \ldots, x_n)$ is ***outside of*** W if $n \geq 2$ and $x_1, \ldots, x_{n-1} \in V \setminus W$, i.e. if every vertex except x_0, x_n lie outside of W. In particular, there exists $1 \leq i \leq n - 1$ such that $x_i \notin W$.

Lemma 2.67. *Let $G = (V, c)$ be a graph which contains no incomplete cycles, W a maximal clique of G and $x, y \in W$, $x \neq y$. Then there exists no simple path $x \to y$ outside of W.*

Proof. Let $(x_0, \ldots, x_n)$ be a simple path $x \to y$ outside of W. Since W is a maximal clique, there exists $1 \leq i \leq n - 1$ and $w \in W$ such that $c(x_i, w) = 0$. Hence, $(y, w, x, x_1, \ldots, x_n)$ is an incomplete cycle in G which is a contradiction. $\qquad\square$

Theorem 2.68. $\mathcal{G}$ *is the class of finite graphs which do not contain any incomplete cycle.*

Proof. By Lemma 2.65, every $G \in \mathcal{G}$ does not contain any incomplete cycle.

For the other direction, let $G = (V, c)$ be any finite graph which does not contain any incomplete cycles and use induction on $|V|$. For $|V| = 1$, we have $G \in \mathcal{G}_0$ because G is complete.

Let $|V| > 1$ and $W = \{w_1, \ldots, w_k\}$ be a maximal clique of G. Furthermore, let $G' = (V, c')$ be the graph which results from G by deleting all edges inside W, i.e.

$$c'(x, y) = \begin{cases} 0 & , x, y \in W \\ c(x, y) & , \text{ otherwise} \end{cases}.$$

Since G is connected, G' has at most k connected components. By Lemma 2.67, for any i, j, there exists no path $w_i \to w_j$ in G outside of W. Hence, there exists no path $w_i \to w_j$ in G'. It follows that G' has exactly k connected components, say $H_1, \ldots, H_k$ where we assume that H_i contains w_i. We can now reconstruct G by gluing each connected component H_i to the maximal clique W at w_i. More precisely, let $G_0 := G \restriction_W$ and for $i > 0$, let

$$G_i := (G_{i-1}, w_i) \oplus (H_i, w_i).$$

Then, $G = G_k$. Furthermore, we have $G_0 = G \restriction_W \in \mathcal{G}_0$ and $H_i \in \mathcal{G}$ by induction since H_i contains no incomplete cycles. Hence, $G \in \mathcal{G}$. $\qquad\square$

Theorem 2.69. (V, d) *is an ERS if and only if G_d contains no incomplete cycles and $(W, d \restriction_W)$ is an ERS for all maximal cliques W of G_d.*

Proof. If (V, d) is an ERS, then for any $W \subseteq V$, $(W, d\restriction_W)$ is an ERS by Proposition 2.26.

Suppose that $(W, d\restriction_W)$ is an ERS for all maximal cliques W of G_d. Then, for each such W, there exists a weighted graph $G_W = (W, c_W)$ such that

$$R_{G_W}(x, y) = d(x, y) \; \forall \, x, y \in W.$$

Let $G_d = (V, c_d)$. Observe that for any two adjacent vertices $x, y \in V$, there exists a unique maximal clique $W(x, y)$ which contains both vertices. Let $G = (V, c)$ where

$$c(x, y) := \begin{cases} 0 & , c_d(x, y) = 0 \\ c_{W(x,y)}(x, y) & , c_d(x, y) > 0 \end{cases}.$$

We claim that $R_G = d$.

Before we prove this claim, observe that G is connected. Indeed, for every maximal clique W of G_d, the graph G_W is connected by definition. Furthermore, $G\restriction_W$ equals G_W and if we interpret a maximal clique as a set of edges, the set of all maximal cliques is a partitioning of all edges of G_d. Hence, G_d being connected implies that G is connected.

Consider $x, y \in V$ such that $c_d(x, y) > 0$. We will see that

$$R_G(x, y) = R_{G\restriction_{W(x,y)}}(x, y) = R_{G_{W(x,y)}}(x, y) = d(x, y)$$

where the second equation is due to $G\restriction_{W(x,y)} = G_{W(x,y)}$ and the third holds by definition of $G_{W(x,y)}$. In order to see that the first equation holds, observe that for any $v, w \in W(x, y)$ there cannot be a path $v \to_G w$ outside of $W(x, y)$. Indeed, suppose there exists such a path. By Lemma 2.52, $c(v', w') > 0$ implies $c_d(v', w') > 0$ for all $v', w' \in V$ and it follows that this path also exists in G_d. But since $W(x, y)$ is a maximal clique in G_d, this would result in an incomplete cycle in G_d which is a contradiction. By Proposition 2.12, it follows that $R_G(x, y) = R_{G\restriction_{W(x,y)}}(x, y)$.

Now consider $x, y \in V$ such that $c_d(x, y) = 0$ and let $p = (x_0, \ldots, x_n)$ be a shortest path $x \to y$ in G_d. We then have

$$d(x, y) = \sum_{i=0}^{n-1} d(x_i, x_{i+1}).$$

Furthermore, for any $1 \leq i \leq n - 1$, all paths $x \to y$ in in G_d use x_i. Indeed, suppose there exists a path $x \to y$ in G_d which does not use x_i, say $q = (y_0, \ldots, y_m)$. This results in a cycle $(x_k, \ldots, x_i, \ldots, x_l, y_{l'+1}, \ldots, y_{k'})$ where $x_k = y_{k'}$ is the last time p and q intersect before p visits x_i and $x_l = y_{l'}$ is the first time p and q intersect after p visited x_i. Since G_d contains no incomplete cycles, it follows that $c(x_k, x_l) > 0$ which is a contradiction to p being a shortest path $x \to y$.

Since $c(v, w) > 0$ implies $c_d(v, w) > 0$ for all $v, w \in V$, it follows that for any $1 \leq i \leq n - 1$, all paths $x \to y$ in G use x_i. Hence, by Proposition 2.27, we have

$$R_G(x, y) = \sum_{i=0}^{n-1} R_G(x_i, x_{i+1}) = \sum_{i=0}^{n-1} d(x_i, x_{i+1}) = d(x, y)$$

where the second equation is due to the fact that $c_d(x_i, x_{i+1}) > 0$ in which case we have already proven that $R_G(x_i, x_{i+1}) = d(x_i, x_{i+1})$. $\qquad\square$

Chapter 3

Infinite Networks

In this chapter, we review the classical theory of electric currents and effective resistances on infinite networks. Most results can be found (in a possibly slightly different form) in the comprehensive textbooks [30] and [36]. We provide some examples in order to develop a deeper understanding of these results.

Throughout this chapter, let $G = (V, c)$ be an infinite graph with Laplacian Δ and energy form $\mathcal{E}$. Furthermore, let $(V_n)_{n \in \mathbb{N}}$ be a *finite exhaustion*. We also assume that V_n is given in such a way that $G_n := G \restriction_{V_n}$ is connected for all n.

3.1 Non-constant harmonic functions

On a finite graph, all harmonic functions are constant. This can be extended to infinite graphs in the sense that all harmonic functions in $l^2(V)$ are constant, see Proposition 1.12. However, in general, there are non-constant harmonic functions in $\mathrm{dom}\,\mathcal{E}$, as the following example shows.

Example 3.1. Consider the network $(\mathbb{Z}, c)$ where

$$c(x, y) = \begin{cases} 2^{\min(|x|,|y|)} & , \ |x - y| = 1 \\ 0 & , \ \text{otherwise} \end{cases}$$

as depicted in Figure 3.1. The function

$$f(z) = \mathrm{sign}(z) \cdot 2(1 - 2^{-|z|}) \tag{3.1}$$

is not constant, harmonic on $(\mathbb{Z}, c)$ and has finite energy. Indeed, since $f(0) = 0$ and $f(-z) = -f(z)$, we have

$$\mathcal{E}(f) = 2 \sum_{n=0}^{\infty} c(n, n+1) \cdot (f(n+1) - f(n))^2 = 2 \sum_{n=0}^{\infty} 2^n \cdot 4 \cdot (2^{-n-1})^2 = 2 \sum_{n=0}^{\infty} 2^{-n} = 4.$$

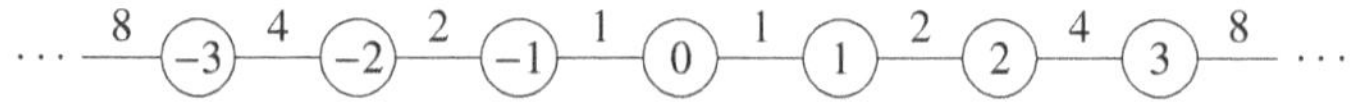

$$\cdots \ \overset{8}{\underline{\quad}} \ \boxed{-3} \ \overset{4}{\underline{\quad}} \ \boxed{-2} \ \overset{2}{\underline{\quad}} \ \boxed{-1} \ \overset{1}{\underline{\quad}} \ \boxed{0} \ \overset{1}{\underline{\quad}} \ \boxed{1} \ \overset{2}{\underline{\quad}} \ \boxed{2} \ \overset{4}{\underline{\quad}} \ \boxed{3} \ \overset{8}{\underline{\quad}} \ \cdots$$

Figure 3.1: A transient network $(\mathbb{Z}, c)$.

Furthermore,

$$(\Delta f)(0) = f(0) - \frac{1}{2}f(-1) - \frac{1}{2}f(1) = 0 - \frac{1}{2} + \frac{1}{2} = 0$$

and, for $n > 1$, we have

$$\frac{1}{3}f(n-1) + \frac{2}{3}f(n+1) = \frac{2}{3}(1 - 2^{-n+1}) + \frac{2}{3} \cdot 2 \cdot (1 - 2^{-n-1})$$
$$= \frac{2}{3}(3 - 3 \cdot 2^{-n})$$
$$= 2(1 - 2^{-n}) = f(n).$$

Hence,

$$(\Delta f)(n) = f(n) - \frac{1}{3}f(n-1) - \frac{2}{3}f(n+1) = 0$$

and

$$(\Delta f)(-n) = f(-n) - \frac{1}{3}f(-(n-1)) - \frac{2}{3}f(-(n+1)) = -(\Delta f)(n) = 0.$$

We denote by $\mathrm{Harm} := \{h \in \mathrm{dom}\,\mathcal{E} \mid \Delta h \equiv 0\}$ the **space of harmonic functions of finite energy**. As we have seen, Harm of an infinite graph may contain more than just the constant functions $\mathbb{R}$. Hence, we can't define effective resistances as on finite graphs because (D) may not have a unique solution.

This leads to fact that one may define several different effective resistances on an infinite graph. The most prominent are the *free* and **wired** effective resistance which we will review in the following sections.

3.2 Free Effective Resistance

Lemma 3.2. *For all $x, y \in V$, $\lim_{n\to\infty} R_{G_n}(x, y)$ exists and is independent of the chosen exhaustion.*

Proof. Let $x, y \in V$. If $x = y$, then $R_{G_n}(x, y) = 0$ for all $n \in \mathbb{N}$ and the limit exists.

If $x \neq y$, let $\phi_n \in l(V_n)$ be the potential of the unit current from x to y in G_n and $\psi_n := \phi_n/\mathcal{E}_{U_n}(\phi_n)$. Then, ψ_n minimizes $\mathcal{E}_{G_n}$ in $\{f \in l(V_n) \mid f(x) = 1, f(y) = 0\}$, see Proposition 2.14. Hence,

$$R_{G_n}(x, y)^{-1} = \mathcal{E}_{G_n}(\psi_n) \leq \mathcal{E}_{G_n}(\psi_{n+1}\restriction_{V_n}) \leq \mathcal{E}_{G_{n+1}}(\psi_{n+1}) \leq R_{G_{n+1}}(x, y)^{-1}.$$

It follows that $R_{G_n}(x, y)$ is monotonically decreasing in n and bounded from below by zero, implying that its limit exists.

The independence regarding the choice of $(V_n)_{n\in\mathbb{N}}$ is a generic argument. For any two exhaustions $(V_n)_{n\in\mathbb{N}}$ and $(V'_n)_{n\in\mathbb{N}}$ with limits $R(x, y)$ and $R'(x, y)$, there exists an exhaustion which contains infinitely many members of $(V_n)_{n\in\mathbb{N}}$ as well as of $(V'_n)_{n\in\mathbb{N}}$. Since the limit exists for this new exhaustion, $R(x, y) = R'(x, y)$ follows. $\qquad\square$

Definition 3.3. For $x, y \in V$, the *free effective resistance* $R^F(x, y)$ of G is defined by

$$R^F(x, y) = \lim_{n \to \infty} R_{G_n}(x, y).$$

Example 3.4. We compute the free effective resistance of $(\mathbb{Z}, c)$ as defined in Example 3.1, see also Figure 3.1. Let $V_n = \{-n, \ldots, 0, \ldots, n\}$. Since G_n is a tree, there exists a unique simple path $\gamma_{xy} : x \to y$ in G. By Proposition 2.30, the effective resistance equals the geodesic metric of G_n, namely

$$R_{G_n}(x, y) = r(\gamma_{xy}) = \sum_{k=\min(x,y)}^{\max(x,y)-1} r(k, k+1) = \sum_{k=\min(x,y)}^{\max(x,y)-1} 2^{-\min(|k|,|k+1|)}$$

for all $x, y \in V_n$. Since this is independent of n, we have

$$R^F(x, y) = \sum_{k=\min(x,y)}^{\max(x,y)-1} 2^{-\min(|k|,|k+1|)}.$$

In particular,

$$R^F(-n, n) = 2 \cdot \sum_{k=0}^{n-1} 2^{-k} = 2 \cdot \frac{1 - 2^{-n}}{1 - 2^{-1}} = 4 \cdot (1 - 2^{-n})$$

which implies that $\left\{R^F(x, y)\right\}_{x, y \in \mathbb{Z}}$ is bounded with diameter 4. In fact, $(\mathbb{Z}, R^F)$ can be embedded isometrically into $([-2, 2], |\cdot|)$ via $z \mapsto \mathrm{sign}(z) \cdot R^F(0, z)$.

Lemma 3.5. *If G is a tree, then $R^F = d_G$.*

Proof. Let $n \in \mathbb{N}$ and $x, y \in V_n$, $x \neq y$. If G is a tree, then so is G_n and there exists exactly one simple path from x to y in G. Since G_n is connected, this path is contained in V_n and $d_{G_n} = d_G \restriction_{V_n}$ follows. Hence,

$$R^F(x, y) = \lim_{n \to \infty} R_{G_n}(x, y) = \lim_{n \to \infty} d_{G_n}(x, y) = d_G(x, y).$$

$\square$

3.3   Wired Effective Resistance

Definition 3.6 (Wired exhaustion). For a given finite exhaustion $(V_n)_{n \in \mathbb{N}}$, we define the associated *wired exhaustion* $(G_n^W)_{n \in \mathbb{N}}$ as follows.

For $n \in \mathbb{N}$, define V_n^W by adding a distinct ∞-vertex to each V_n, namely

$$V_n^W := V_n \,\dot{\cup}\, \{v_\infty\}. \tag{3.2}$$

Furthermore, let $G_n^W := (V_n^W, c_n^W)$ where $c_n^W \restriction_{V_n} = c \restriction_{V_n}$ and, for $x \in V_n$,

$$c_n^W(x, v_\infty) = c(x, V \setminus V_n). \tag{3.3}$$

We denote by $\mathcal{E}_n^W$, Δ_n^W and R_n^W the associated energy form, Laplacian and effective resistance of G_n^W, respectively.

Remark 3.7.

1. Note that the term *finite exhaustion* refers to a sequence of vertex sets while *wired exhaustion* always refers to a sequence of graphs.

2. For any $x \in V_n$, we have $(c_n^W)_x = \sum_{y \in V_n^W} c_n^W(x, y) = c_x$.

For $A \subseteq V$, define

$$l_{\mathrm{fin}}(V; A) := \left\{ f \in l(V) \mid \exists\, r \in \mathbb{R} : f\!\restriction_{V \setminus A} \equiv r \right\} \tag{3.4}$$

and, for $f \in l(V_n^W)$,

$$L_n^W(f) := \mathbf{1}_{V_n} \cdot f + \mathbf{1}_{V \setminus V_n} \cdot f(v_\infty). \tag{3.5}$$

Then, L_n^W is a bijection $l(V_n^W) \to l_{\mathrm{fin}}(V; V_n)$ satisfying

$$\mathcal{E}_n^W(f) = \mathcal{E}(L_n^W(f)). \tag{3.6}$$

Lemma 3.8. *For all $x, y \in V$, $\lim_{n \to \infty} R_n^W(x, y) \le R^F(x, y)$ exists and is independent of the chosen exhaustion.*

Proof. Fix $x, y \in V$. If $x = y$, then $R_n^W(x, y) = 0$ for all n and its limit exists.

If $x \neq y$, let ϕ_n be the potential of the unit current from x to y in G_n^W and $\psi_n := \phi_n / \mathcal{E}_n^W(\phi^n)$ its associated minimizer of $\mathcal{E}_n^W$. Then, $L_n^W(\psi_n) \in l_{\mathrm{fin}}(V; V_n) \subseteq l_{\mathrm{fin}}(V; V_{n+1})$. Hence,

$$f_{n+1} := (L_{n+1}^W)^{-1}(L_n^W(\psi_n)) \in l(V_{n+1}^W)$$

exists and satisfies $f_{n+1}(x) = 1$, $f_{n+1}(y) = 0$. It follows that

$$R_n^W(x, y)^{-1} = \mathcal{E}_n^W(\psi_n) = \mathcal{E}(L_n^W(\psi_n)) = \mathcal{E}_{n+1}^W(f_{n+1}) \ge \mathcal{E}_{n+1}^W(\psi_{n+1}) = R_{n+1}^W(x, y)^{-1}.$$

Hence, $R_n^W(x, y)$ is monotonically increasing. Since

$$R_n^W(x, y)^{-1} = \mathcal{E}_n^W(\psi_n) \ge \mathcal{E}_{G_n}(\psi_n) \ge R_{G_n}(x, y)^{-1}$$

we have $R_n^W(x, y) \le R_{G_n}(x, y)$ for all n. As seen in the proof of Lemma 3.2, $R_{G_n}(x, y)$ in a decreasing sequence and it follows that

$$\lim_{n \to \infty} R_n^W(x, y) \le \lim_{n \to \infty} R_{G_n}(x, y) = R^F(x, y)$$

exists and is finite.

That the limit is independent of the choice of $(V_n)_{n \in \mathbb{N}}$ follows by the same generic argument as in the proof of Lemma 3.2. $\square$

Definition 3.9 (Wired effective resistance). For $x, y \in V$, the *wired effective resistance* $R^W(x, y)$ of G is defined by

$$R^W(x, y) := \lim_{n \to \infty} R_n^W(x, y). \tag{3.7}$$

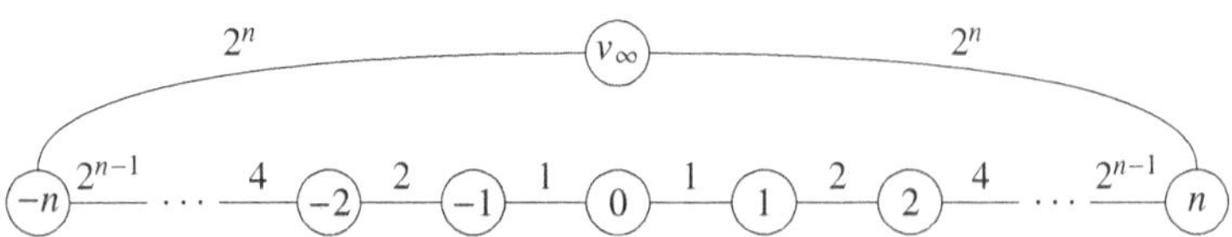

Figure 3.2: The graph G_n^W associated to a wired exhaustion of $\mathbb{Z}$.

Example 3.10. Continuing our example from before (Examples 3.1 and 3.4), we will compute R^W of $(\mathbb{Z}, c)$ as shown in Figure 3.1. For $n \in \mathbb{N}$, let $V_n = \{-n, \ldots, 0, \ldots, n\}$ and $V_n^W := V_n \,\dot{\cup}\, \{v_\infty\}$. Then, the definition of G_n^W yields the graph depicted in Figure 3.2. Applying the Star-Mesh transform (Theorem 2.6) to v_∞, yields a cycle graph $G'_n = (V_n, c'_n)$ where

$$c'_n(-n, n) = c'_n(n, -n) = 2^{n-1}$$

and $c'_n(x, y) = c(x, y)$ for all other $x, y \in V_n$.

Fix $x, y \in V, x < y$. Then, Example 2.10 states that

$$R_n^W(x, y) = \left(\frac{1}{r(\gamma_1)} + \frac{1}{r(\gamma_2)} \right)^{-1}$$

if γ_1, γ_2 are the two distinct simple paths $x \to y$ in G'_n. Let γ_1 be the path which does not use the edge $(-n, n)$. Then, $r(\gamma_1) = R^F(x, y)$ as seen in Example 3.4 and

$$r(\gamma_2) = R^F(x, -n) + 2^{-n+1} + R^F(n, y) = 2^{-n+1} + R^F(-n, n) - R^F(x, y).$$

By Examples 3.4, we have $R^F(-n, n) \to 4$ and it follows that

$$\begin{aligned}
R^W(x, y) &= \lim_{n \to \infty} \left(\frac{1}{R^F(x, y)} + \frac{1}{2^{-n+1} + R^F(-n, n) - R^F(x, y)} \right)^{-1} \\
&= \frac{1}{4} R^F(x, y)(4 - R^F(x, y)).
\end{aligned}$$

In particular, $R^W \neq R^F$, $R^W(x, y) \leq 2$ and $\lim_{n \to \infty} R^W(-n, n) \to 0$.

Using the notion of wired exhaustions, we can define the effective resistance between a vertex and infinity. This will be very useful when investigating whether the network is recurrent or transient.

Definition 3.11 (Effective resistance to infinity). For $x \in V$, we define the *effective resistance between x and infinity* as

$$R(x, \infty) := \lim_{n \to \infty} R_n^W(x, v_\infty) \tag{3.8}$$

if the right-hand side exists.

Proposition 3.12. *For every $x \in V$, $R(x, \infty)$ is either real number or $+\infty$. Furthermore, we have $R(x, \infty) < \infty$ if and only if G is transient.*

Proof. We denote by $\mathbf{P}_x$ and $\mathbf{P}_x^{W,n}$ the distributions of the random walks on G and G_n^W, respectively. Then, the transition probabilities of $\mathbf{P}_x$ and and $\mathbf{P}_x^{W,n}$ agree on V_n, i.e.

$$\mathbf{P}_x^{W,n}[X_{k+1} = z \mid X_k = y] = \mathbf{P}_x[X_{k+1} = z \mid X_k = y]$$

for all $y, z \in V_n$. Furthermore, in G_n^W the probability of stepping to v_∞ corresponds to the probability of stepping out of V_n in G

$$\mathbf{P}_x^{W,n}[X_{k+1} = v_\infty \mid X_k = y] = \mathbf{P}_x[X_{k+1} \notin V_n \mid X_k = y].$$

Hence, $\mathbf{P}_x^{W,n}[\tau_{v_\infty} \leq \tau_x^+] = \mathbf{P}_x[\tau_{V \setminus V_n} \leq \tau_x^+]$. It follows by Proposition 2.19 that

$$\lim_{n \to \infty} R_n^W(x, v_\infty) = \lim_{n \to \infty} \frac{1}{(c_n^W)_x \cdot \mathbf{P}_x^{W,n}[\tau_{v_\infty} \leq \tau_x^+]}$$
$$= \lim_{n \to \infty} \frac{1}{c_x \cdot \mathbf{P}_x[\tau_{V \setminus V_n} \leq \tau_x^+]} = \frac{1}{c_x \cdot \mathbf{P}_x[\tau_x^+ = \infty]} .$$

The last equality follows from the Monotone Convergence Theorem and the fact that $\mathbf{1}_{\{\tau_x^+ < \tau_{V \setminus V_n}\}} \nearrow \mathbf{1}_{\{\tau_x^+ < \infty\}}$ almost everywhere.

Hence, $R(x, \infty) := \lim_{n \to \infty} R_n^W(x, v_\infty)$ exists in $\mathbb{R}$ or is $+\infty$. Furthermore, $R(x, \infty) < \infty$ if and only if $\mathbf{P}_x[\tau_x^+ = \infty] > 0$ which is the definition of transience. $\square$

Example 3.13. Based on our examples of $(\mathbb{Z}, c)$, we can explicitly compute the effective resistance between 0 and ∞. As seen in Example 3.10, we have

$$R_n^W(0, v_\infty) = \frac{r(\gamma_n)}{2} = \frac{1}{2} \cdot R^F(0, n+1).$$

where γ_n is the path $(0, 1, \ldots, n, v_\infty)$. Hence,

$$R(0, \infty) = \lim_{n \to \infty} \frac{1}{2} \cdot R^F(0, n+1) = 1$$

and we have confirmed that $(\mathbb{Z}, c)$ is indeed transient.

3.4 Royden decomposition

Recall that we defined the Hilbert spaces $\mathrm{H}_x := (\mathrm{dom}\,\mathcal{E}, \mathcal{E} + \delta_x^2)$ and $\mathcal{H}_\mathcal{E} := (\mathrm{dom}\,\mathcal{E}/\mathbb{R}, \mathcal{E})$ in Section 1.4. Let $l_{\mathrm{fin}}(V) := \{f \in l(V) \mid |\,\mathrm{supp}\,f| < \infty\}$ be the *space of finitely supported functions*.

Lemma 3.14. *Every function with finite support has finite energy, i.e. $l_{\mathrm{fin}}(V) \subseteq \mathrm{dom}\,\mathcal{E}$.*

Proof. Let $f \in l_{\mathrm{fin}}(V)$ and $S = \mathrm{supp}\,f$. Then, $M := \max_{x \in V} |f(x)| = \max_{x \in S} |f(x)|$ exists and

$$c(S) := \sum_{\substack{x \in S \\ y \in V}} c(x, y) = \sum_{x \in S} c_x < \infty.$$

Hence,

$$\mathcal{E}(f) = \frac{1}{2} \sum_{x,y \in V} c(x,y) \cdot (f(x) - f(y))^2$$

$$\leq \frac{1}{2} \sum_{x,y \in S} c(x,y) \cdot 4M^2 + \sum_{\substack{x \in S \\ y \notin S}} c(x,y) \cdot M^2 \leq 3M^2 c(S) < \infty$$

$\square$

For any $x \in V$, let $l_0(V)$ be the closure of $l_{\mathrm{fin}}(V)$ in H_x. Since the topology of H_x is independent of x, so is the definition of $l_0(V)$.

Lemma 3.15. *The space $l^2(V)$ is a dense subspace of $l_0(V)$. However, in general $l^2(V) \neq l_0(V)$ holds.*

Proof. Let $f \in l^2(V)$ and $(V_n)_{n \in \mathbb{N}}$ be an exhaustion of V. For $f_n := f \cdot \mathbf{1}_{V_n}$, we have $f_n \in l_{\mathrm{fin}}(V)$, $f_n(x) \to f(x)$ for all $x \in V$ and

$$\mathcal{E}(f - f_n) = \|\nabla(f - f_n)\|_E^2 \leq 2 \cdot \|f - f_n\|_V^2 \to 0$$

by Proposition 1.11. Hence, $\|f - f_n\|_{H_x} \to 0$ and $f \in l_0(V)$ follows. $l^2(V)$ is dense in $l_0(V)$ because it contains $l_{\mathrm{fin}}(V)$ which is dense.

To see that $l^2(V) \neq l_0(V)$ in general, consider the unweighted graph G with vertex set $\mathbb{N}$ and $(x,y) \in E(G)$ iff $|x - y| = 1$. For $f(n) := 1/\sqrt{n}$, we have

$$\|f\|_V^2 = \sum_{n \in \mathbb{N}} c_n f(n)^2 \geq \sum_{n \in \mathbb{N}} \frac{1}{n} = \infty$$

and thus $f \notin l^2(V)$. To see that $f \in l_0(V)$, let $V_n := \{1, \ldots, n\}$ and $f_n := f \cdot \mathbf{1}_{V_n} \in l_{\mathrm{fin}}(V)$. It follows that $f_n(x) \to f(x)$ for all $x \in \mathbb{N}$ and

$$\mathcal{E}(f - f_n) = \sum_{m \in \mathbb{N}} ((f - f_n)(m) - (f - f_n)(m + 1))^2$$

$$= f(n + 1)^2 + \sum_{m \geq n} (f(m) - f(m + 1))^2$$

$$= \frac{1}{n + 1} + \sum_{m \geq n} \left(\frac{1}{\sqrt{m}} - \frac{1}{\sqrt{m + 1}} \right)^2$$

$$= \frac{1}{n + 1} + \sum_{m \geq n} \left(\frac{1}{\sqrt{m(m + 1)} \cdot (\sqrt{m} + \sqrt{m + 1})} \right)^2$$

$$\leq \frac{1}{n + 1} + \sum_{m \geq n} \frac{1}{m^2} \to 0.$$

Hence, $\|f - f_n\|_{H_x} \to 0$ and thus $f \in l_0(V) \setminus l^2(V)$. $\square$

Definition 3.16. Let

- Fin $= l_0(V) + \mathbb{R}$,

- $\mathcal{F}in := \mathrm{Fin}/\mathbb{R}$ and

- $\mathcal{H}arm := \mathrm{Harm}/\mathbb{R}$.

Note that Fin is a subspace of H_x while $\mathcal{F}in$ and $\mathcal{H}arm$ are subspaces of $\mathcal{H}_\mathcal{E}$.

Proposition 3.17. *We have*

$$\mathrm{Harm} = \{h \in \mathrm{dom}\,\mathcal{E} \mid \mathcal{E}(f,h) = 0 \;\forall\; f \in \mathrm{Fin}\} \tag{3.9}$$

and

$$\mathrm{Fin} \cap \mathrm{Harm} = \mathbb{R}. \tag{3.10}$$

Proof. Let $h \in \mathrm{Harm}$, $f \in l_0(V)$ and $f_n \in l_{\mathrm{fin}}(V)$ such that $f_n \to f$ in H_x. In particular, $f \in l^2(V)$ and $\mathcal{E}(f_n - f) \to 0$. Hence, by Proposition 1.14, we have

$$\mathcal{E}(f,h) = \lim_{n\to\infty} \mathcal{E}(f_n, h) = \lim_{n\to\infty} \langle f_n, \Delta h \rangle_V = 0.$$

Conversely, let $h \in \mathrm{dom}\,\mathcal{E}$ such that $\mathcal{E}(f,h) = 0$ for all $f \in \mathrm{Fin}$. For $v \in V$, $\mathbf{1}_v \in \mathrm{Fin} \cap l^2(V)$ and, by Proposition 1.14, we have

$$0 = \mathcal{E}(\mathbf{1}_v, h) = \langle \mathbf{1}_v, \Delta h \rangle_V = c_v \cdot (\Delta h)(v).$$

Hence, $h \in \mathrm{Harm}$. If $h \in \mathrm{Harm} \cap \mathrm{Fin}$, we have $\mathcal{E}(h) = \mathcal{E}(h,h) = 0$ and it follows that h is constant. $\square$

The following result (as well as Corollary 3.21) is a discrete version of Royden's decomposition theorem on Riemannian surfaces [33]. As such, the literature refers to it as the ***Royden decomposition*** (see [19, 30, 36]).

Theorem 3.18. *(Royden decomposition)* $\mathcal{H}_\mathcal{E} = \mathcal{F}in \oplus \mathcal{H}arm$.

Proof. Since $\langle [f], [g] \rangle_{\mathcal{H}_\mathcal{E}} = \mathcal{E}(f,g)$ for all $[f], [g] \in \mathcal{H}_\mathcal{E}$, Proposition 3.17 implies that $\mathcal{H}arm$ is the orthogonal complement $\mathcal{F}in^\perp$ of $\mathcal{F}in$ and the claim follows. $\square$

Proposition 3.19. *G is recurrent if and only if $\mathbf{1} \in l_0(V)$.*

Proof. Since $l_0(V)$ is a closed subspace of H_x, we may compute the distance between $\mathbf{1}$ and $l_0(V)$ as follows.

$$\|\mathbf{1} - l_0(V)\|_{\mathrm{H}_x}^2 = \min_{f \in l_0(V)} \|\mathbf{1} - f\|_{\mathrm{H}_x}^2 = \min_{f \in l_0(V)} \left[(1 - f(x))^2 + \mathcal{E}(1 - f) \right]$$

$$= \min_{f \in D} \left[(1 - f(x))^2 + \mathcal{E}(f) \right]$$

Since $l_0(V)$ is a linear space and $\mathcal{E}(f + k) = \mathcal{E}(f)$ for any $k \in \mathbb{R}$, the minimizer of the above problem must satisfy $f(x) = 1$. Hence, we have

$$\|\mathbf{1} - l_0(V)\|_{\mathrm{H}_x}^2 = \min_{\substack{f \in l_0(V) \\ f(x)=1}} \mathcal{E}(f). \tag{3.11}$$

Suppose that $\psi \in l_0(V)$ is such a minimizer. By definition of $l_0(V)$, there exist $\psi_n \in l_{\mathrm{fin}}(V)$ such that $\psi_n \to \psi$ point-wise and w.r.t. $\mathcal{E}$. Passing to a subsequence of $(V_n)_{n \in \mathbb{N}}$, we may assume that $\operatorname{supp} \psi_n \subseteq V_n$. Define $\widetilde{\psi}_n : V \to \mathbb{R}$ by

$$\widetilde{\psi}_n(v) := \frac{\psi_n(v)}{\psi_n(x)} \, , \; v \in V.$$

For sufficiently large n, this is well-defined because $\psi_n(x) \to \psi(x) = 1$. Then, $\widetilde{\psi}_n(x) = 1$, $\widetilde{\psi}_n \restriction_{V \setminus V_n} \equiv 0$ and

$$\lim_{n \to \infty} \mathcal{E}(\widetilde{\psi}_n) = \lim_{n \to \infty} \frac{1}{\psi_n(x)^2} \mathcal{E}(\psi_n) = \mathcal{E}(\psi).$$

It follows that

$$\min_{\substack{f \in l_0(V) \\ f(x)=1}} \mathcal{E}(f) \leq \lim_{n \to \infty} \min_{\substack{f \in l_0(V) \\ f(x)=1 \\ f \restriction_{V \setminus V_n} \equiv 0}} \mathcal{E}(f) \leq \lim_{n \to \infty} \mathcal{E}(\widetilde{\psi}_n) = \mathcal{E}(\psi) = \min_{\substack{f \in l_0(V) \\ f(x)=1}} \mathcal{E}(f). \tag{3.12}$$

Recall the bijection $L_n^W : l(V_n^W) \to l_{\mathrm{fin}}(V; V_n)$ from Section 3.3. It satisfies $\mathcal{E}_n^W(f) = \mathcal{E}(L_n^W(f))$ for any $f \in l(V_n^W)$. Hence, we have

$$\|\mathbf{1} - l_0(V)\|_{\mathrm{H}_x}^2 = \lim_{n \to \infty} \min_{\substack{f \in l_0(V) \\ f(x)=1 \\ f \restriction_{V \setminus V_n} \equiv 0}} \mathcal{E}(f) = \lim_{n \to \infty} \min_{\substack{f \in l(V_n^W) \\ f(x)=1 \\ f(v_\infty)=0}} \mathcal{E}_n^W(f)$$

$$= \lim_{n \to \infty} R_n^W(x, v_\infty)^{-1} = R(x, \infty)^{-1}.$$

by Corollary 2.15. It follows that $\|\mathbf{1} - l_0(V)\|_{\mathrm{H}_x} > 0$ if and only if $R(x, \infty) < \infty$ which is equivalent to G being transient by Proposition 3.12. The claim follows. $\qquad\square$

Corollary 3.20. *G is recurrent if and only if* $\operatorname{dom} \mathcal{E} = l_0(V)$. *If G is transient,* $|[f] \cap l_0(V)| = 1$ *for all $f \in$ Fin.*

Proof. If $\operatorname{dom} \mathcal{E} = l_0(V)$, then $\mathbf{1} \in l_0(V)$ and G is recurrent by Proposition 3.19. If G is recurrent, we have $\mathbf{1} \in l_0(V)$. In particular, there are $f_n \in l_{\mathrm{fin}}(V)$ such that $f_n \to \mathbf{1}$ in H_x. Let $f_n' = m(f_n) = \max(0, \min(f_n, 1)) \in l_{\mathrm{fin}}(V)$. Then, $m(f_n) \to \mathbf{1}$ point-wise and

$$\mathcal{E}(f_n' - \mathbf{1}) = \mathcal{E}(m(f_n)) \leq \mathcal{E}(f_n) = \mathcal{E}(f_n - \mathbf{1}) \to 0.$$

Hence, $f_n' \to \mathbf{1}$ in H_x and $f_n'(V) \subseteq [0, 1]$. Let $g \in \operatorname{dom} \mathcal{E}$ be bounded, say $|g| \leq K$. Then, we have $f_n' \cdot g \in l_{\mathrm{fin}}(V)$, $f_n' \cdot g \to g$ point-wise and

$$(f_n(v)g(v) - f_n(w)g(w))^2 = (f_n(v)(g(v) - g(w)) + (f_n(v) - f_n(w))g(w))^2$$

$$\leq 2(g(v) - g(w))^2 + 2K(f_n(v) - f_n(w))^2$$

for every $v, w \in V$. Hence, we may apply Lebesgue's Dominated Convergence Theorem to obtain $f'_n \cdot g \to g$ in H_x and thus $g \in l_0(V)$. By Lemma 1.18, the bounded functions are dense in H_x and $l_0(V) = \operatorname{dom} \mathcal{E}$ follows.

If G is transient, we have $l_0(V) \cap \mathbb{R} = \{0\}$. For any $f \in \operatorname{Fin}$, let $f_1, f_2 \in [f] \cap l_0(V)$. Then, $f_1 - f_2$ is constant and thus zero. $\qquad\square$

Corollary 3.21. *We have*

$$\operatorname{dom} \mathcal{E} = l_0(V) + \operatorname{Harm}. \tag{3.13}$$

If G is transient, this decomposition is unique.

Proof. Let $\phi \in \operatorname{dom} \mathcal{E}$. By Theorem 3.18, there exist unique $[f] \in \mathcal{F}in, [h] \in \mathcal{H}arm$ such that $[\phi] = [f] + [h]$. Since $\mathcal{F}in = (l_0(V) + \mathbb{R})/\mathbb{R}$, we may choose a representative f_0 of $[f]$ which is in $l_0(V)$. Then, there exists $k \in \mathbb{R}$ such that $\phi + k = f_0 + h$. For $h_0 := h - k$, we have $h_0 \in \operatorname{Harm}$ and $\phi = f_0 + h_0$ which is the desired decomposition.

Now assume that G is transient and let $f_0, f_1 \in l_0(V)$, $h_0, h_1 \in \operatorname{Harm}$ such that

$$f_0 + h_0 = \phi = f_1 + h_1.$$

It follows from the Royden Decomposition (Theorem 3.18) that $f_0, f_1 \in [f] \cap l_0(V)$ for some $f \in \operatorname{Fin}$. By Corollary 3.20, transience of G implies $f_0 = f_1$. Hence, $h_0 = h_1$ follows also.1

$\qquad\square$

Remark 3.22. The decomposition (3.13) is never orthogonal with respect to the inner product of x. Indeed, we have $\mathbf{1}_x \in l_0(V)$, $\mathbf{1} \in \operatorname{Harm}$ and $\langle \mathbf{1}_x, \mathbf{1} \rangle_{H_x} = \mathcal{E}(\mathbf{1}_x, \mathbf{1}) + 1 = 1$.

Chapter 4

Resistance Forms and Metrics

In the previous chapter, we have defined effective resistances on infinite graphs by taking limits of effective resistances of finite subgraphs. We will now look at an alternative approach developed by Kigami in [23, 24], namely the theory of resistance forms and metrics. This will allow us to define resistance metrics on arbitrary sets without the need of an underlying graph structure.

We start by reproducing essential definitions and statements from [20, 23, 24], which are the fundamental inspiration of this chapter. Afterwards, we will show that our characterization of finite effective resistance spaces (Theorem 2.44) can be extended to more general resistance metrics, see Theorem 4.32.

Definition 4.1 (Resistance form). Let X be a set. A pair $(\mathcal{F}, D[\mathcal{F}])$ of a linear subspace $D[\mathcal{F}] \subseteq l(X)$ and a non-negative symmetric quadratic form $\mathcal{F} : D[\mathcal{F}] \to \mathbb{R}$ is called a *resistance form on* X if the following hold:

(RF-1) $\mathbf{1} \in D[\mathcal{F}]$ and $\mathcal{F}(f) = 0$ if and only if f is constant on X.

(RF-2) For any $x \in X$, $(D[\mathcal{F}], \mathcal{F} + \delta_x^2)$ is a Hilbert space.

(RF-3) For any finite subset $F \subseteq X$ and any $f \in l(F)$, there exists $g \in D[\mathcal{F}]$ such that $g \restriction_F = f$.

(RF-4) For any $x, y \in X, x \neq y$, we have

$$0 < R_{(\mathcal{F}, D[\mathcal{F}])}(x, y) := \sup \left\{ \frac{|f(x) - f(y)|^2}{\mathcal{F}(f)} \mid f \in D[\mathcal{F}], \mathcal{F}(f) > 0 \right\} < \infty.$$

(RF-5) If $f \in D[\mathcal{F}]$, then $m(f) \in D[\mathcal{F}]$ where $m(f)(x) := \min(1, \max(0, f(x)))$ and $\mathcal{F}(m(f)) \leq \mathcal{F}(f)$.

If there is no risk of confusion, we will drop the domain and simply refer to $\mathcal{F}$ as a resistance form. In this case we also write $R_{\mathcal{F}}$ instead of $R_{(\mathcal{F}, D[\mathcal{F}])}$.

Remark 4.2. (RF-2) is equivalent to

$$(D[\mathcal{F}]/\mathbb{R}, \mathcal{F}) \text{ is a Hilbert space.} \tag{RF2'}$$

We begin by showing some basic properties of resistance forms.

Lemma 4.3 ([20, Lemma 1.14]). *Let $(\mathcal{F}, D[\mathcal{F}])$ be a resistance form on a set X. Then, for any $x, y \in X, x \neq y$, such that $\mathbf{1}_x, \mathbf{1}_y \in D[\mathcal{F}]$, we have*

$$\mathcal{F}(\mathbf{1}_x, \mathbf{1}_y) \leq 0. \tag{4.1}$$

Proof [20]. For $x \neq y$ and $t > 0$, let $f_t := \mathbf{1}_x - t\mathbf{1}_y$. Then, $m(f_t) = \mathbf{1}_x$ and, by (RF-5),

$$\mathcal{F}(\mathbf{1}_x) = \mathcal{F}(m(f_t)) \leq \mathcal{F}(f_t) = \mathcal{F}(\mathbf{1}_x) - 2t\mathcal{F}(\mathbf{1}_x, \mathbf{1}_y) + t^2\mathcal{F}(\mathbf{1}_y).$$

Hence,

$$\mathcal{F}(\mathbf{1}_x, \mathbf{1}_y) \leq \frac{t}{2}\mathcal{F}(\mathbf{1}_y)$$

for all $t > 0$. Hence, $t \to 0$ yields $\mathcal{F}(\mathbf{1}_x, \mathbf{1}_y) \leq 0$. $\qquad\square$

Lemma 4.4. *If $f_n \to f$ in $(D[\mathcal{F}], \mathcal{F} + \delta_x^2)$ for some $x \in X$, then $f_n \to f$ point-wise on X.*

Proof. Point-wise convergence in x follows directly from $\delta_x^2(f_n - f) \to 0$. Let $y \in X$, $y \neq x$. Then,

$$\lim_{n \to \infty} (f_n(x) - f(x) - (f_n(y) - f(y)))^2 \leq \lim_{n \to \infty} R_{\mathcal{F}}(x, y) \cdot \mathcal{F}(f_n - f) = 0$$

by (RF-4). Hence, $\lim_{n \to \infty} f_n(y) = \lim_{n \to \infty}(f(y) + f_n(x) - f(x)) = f(y)$. $\qquad\square$

Proposition 4.5. *For any $x, y, z \in X$, there exists a unique function $g_z^{xy} \in D[\mathcal{F}]$ such that $g_z^{xy}(z) = 0$ and*

$$\forall f \in D[\mathcal{F}] : \mathcal{E}(f, g_z^{xy}) = f(x) - f(y). \tag{4.2}$$

Proof. Let $f \in D[\mathcal{F}]$, $x, y \in X$ and let $\delta_{xy} : D[\mathcal{F}] \to \mathbb{R}$ be the linear functional defined by $\delta_{xy}f = f(x) - f(y)$. By (RF-4), we have

$$|\delta_{xy}f|^2 = |f(x) - f(y)|^2 \leq R_{\mathcal{F}}(x, y) \cdot \mathcal{F}(f) \leq R_{\mathcal{F}}(x, y) \cdot (\mathcal{F}(f) + f(z)^2)$$

which means that δ_{xy} is bounded. By Riesz theorem, there exits $g_z^{xy} \in D[\mathcal{F}]$ such that

$$f(x) - f(y) = \delta_{xy}f = (\mathcal{F} + \delta_z^2)(f, g_z^{xy}) = \mathcal{F}(f, g_z^{xy}) + f(z)g_z^{xy}(z).$$

For $f = \mathbf{1}$, this yields $g_z^{xy}(z) = \mathbf{1}(x) - \mathbf{1}(y) = 0$. $\qquad\square$

In what follows, we set $\mathcal{F}(f) = \infty$ for all $f \in l(X) \setminus D[\mathcal{F}]$.

Lemma 4.6 ([20, Lemma 1.4])**.** *$\mathcal{F}$ is lower semi-continuous w.r.t. the topology of point-wise convergence. More precisely, if $f_n, f \in l(X)$ such that $f_n \to f$ point-wise, then*

$$\mathcal{F}(f) \leq \liminf_{n \to \infty} \mathcal{F}(f_n).$$

In particular, $f \in D[\mathcal{F}]$ if and only if there exists a sequence $f_n \in D[\mathcal{F}]$ such that $f_n \to f$ point-wise and $\liminf_{n \to \infty} \mathcal{F}(f_n) < \infty$.

Proof. If $\liminf_{n \to \infty} \mathcal{F}(f_n) = \infty$, there is nothing to show. By passing to a subsequence, we may assume that $\lim_{n \to \infty} \mathcal{F}(f_n) = \liminf_{n \to \infty} \mathcal{F}(f_n) < \infty$ and $f_n \in D[\mathcal{F}]$.

Fix $y \in X$. Since convergence in $(D[\mathcal{F}], \mathcal{F} + \delta_y^2)$ implies point-wise convergence, the set $D_y := \{g \in D[\mathcal{F}] \mid g(y) = 0\}$ is a closed subspace. Hence, $(D_y, \mathcal{F})$ is a Hilbert space with norm $\|\cdot\| = \sqrt{\mathcal{F}}$. Let $f_n' := f_n - f_n(y) \in D_y$. Since $\mathcal{F}(f_n') = \mathcal{F}(f_n)$ is bounded, the sequence $(f'_n)_{n\in\mathbb{N}}$ has a weakly convergent subsequence $(f'_{n_k})_{k\in\mathbb{N}}$. More precisely, there exists $f' \in D_y$ such that

$$\lim_{k\to\infty} \mathcal{F}(f'_{n_k}, g) = \mathcal{F}(f', g) \;\forall\, g \in D_y.$$

Let g_{xy}, $x \in X$, be the functions from Proposition 4.5 such that $g_{xy}(y) = 0$. Then, $g_{xy} \in D_y$ and we have

$$f'(x) = \mathcal{F}(f', g_{xy}) = \lim_{k\to\infty} \mathcal{F}(f'_{n_k}, g_{xy}) = \lim_{k\to\infty} f'_{n_k}(x) = f(x) - f(y).$$

It follows that $f - f(y) = f' \in D_y$ and thus $f \in D[\mathcal{F}]$. The norm of Hilbert spaces is weakly lower semi-continuous and it follows that

$$\mathcal{F}(f) = \mathcal{F}(f') = \|f'\|^2 \leq \liminf_{k\to\infty} \left\|f'_{n_k}\right\|^2$$
$$= \liminf_{k\to\infty} \mathcal{F}(f'_{n_k}) = \liminf_{n\to\infty} \mathcal{F}(f_{n_k}) = \liminf_{n\to\infty} \mathcal{F}(f_n).$$

$\square$

4.1   Resistance Forms on Finite Sets

If X is a finite set, resistance forms become a somewhat simpler object. In fact, we will show that every resistance form on a finite set is the energy form of a connected graph. This is not true for resistance forms on countable sets as we will see later.

Lemma 4.7. *Let $(\mathcal{F}, D[\mathcal{F}])$ be a resistance form on a finite set X. Then,*

$$\mathcal{F}(f) = \frac{1}{2} \sum_{x,y\in X} (-\mathcal{F}(\mathbf{1}_x, \mathbf{1}_y)) \cdot (f(x) - f(y))^2. \tag{4.3}$$

Furthermore, for any $x, y \in X$, $x \neq y$, there exist $x_0, \ldots, x_n \in X$ such that $x_0 = x, x_n = y$ and

$$\mathcal{F}(\mathbf{1}_{x_k}, \mathbf{1}_{x_{k+1}}) < 0$$

for all $k = 0, \ldots, n-1$.

Proof. Since X is finite, we have

$$\mathcal{F}(f, g) = \mathcal{F}\left(\sum_{x\in X} f(x)\mathbf{1}_x, \sum_{y\in X} g(y)\mathbf{1}_y\right) = \sum_{x,y\in X} f(x)g(y) \cdot \mathcal{F}(\mathbf{1}_x, \mathbf{1}_y).$$

By (RF-1), we have

$$0 = \mathcal{F}(f^2, 1) = \sum_{x,y\in X} f^2(x) \cdot \mathcal{F}(\mathbf{1}_x, \mathbf{1}_y)$$

and therefore

$$
\begin{aligned}
\mathcal{F}(f) &= -\frac{1}{2}\mathcal{F}(f^2, 1) + \mathcal{F}(f, f) - \frac{1}{2}\mathcal{F}(1, f^2) \\
&= -\frac{1}{2}(\mathcal{F}(f^2, 1) - 2\mathcal{F}(f, f) + \mathcal{F}(1, f^2)) \\
&= -\frac{1}{2}\sum_{x,y\in X}(f^2(x) - 2f(x)f(y) + f^2(y)) \cdot (\mathcal{F}(\mathbf{1}_x, \mathbf{1}_y)) \\
&= \frac{1}{2}\sum_{x,y\in X}(-\mathcal{F}(\mathbf{1}_x, \mathbf{1}_y)) \cdot (f(x) - f(y))^2.
\end{aligned}
$$

For the second claim, assume $|X| \geq 2$, fix $x \in X$ and let

$$
A(x) := \left\{ y \in X \mid \exists\, n \in \mathbb{N}_0\ \exists\, x_1, \ldots, x_n \in X : x_1 = x, x_n = y, \mathcal{F}(\mathbf{1}_{x_i}, \mathbf{1}_{x_{i+1}}) < 0 \right\}.
$$

Then we have

$$
\mathcal{F}(\mathbf{1}_{A(x)}) = \frac{1}{2}\sum_{v\in X}\sum_{w\in X}(-\mathcal{F}(\mathbf{1}_v, \mathbf{1}_w))(\mathbf{1}_{A(x)}(v) - \mathbf{1}_{A(x)}(w))^2 = 0
$$

because $v \in A(x)$, $w \notin A(x)$ implies $\mathcal{F}(\mathbf{1}_v, \mathbf{1}_w) = 0$ by definition of $A(x)$. Hence, (RF-2) implies that $\mathbf{1}_{A(x)}$ is constant on X. Since $|X| \geq 2$, $\mathbf{1}_x$ is not constant on X and it follows that

$$
0 < \mathcal{F}(\mathbf{1}_x) = \mathcal{F}(\mathbf{1}_x, 1) - \sum_{v\neq x}\mathcal{F}(\mathbf{1}_x, \mathbf{1}_v) = -\sum_{v\neq x}\mathcal{F}(\mathbf{1}_x, \mathbf{1}_v).
$$

Hence, there exists $v \neq x$ such that $\mathcal{F}(\mathbf{1}_x, \mathbf{1}_v) < 0$ and it follows that $A(x) \neq \emptyset$. Thus $\mathbf{1}_{A(x)} \equiv 1$ and $A(x) = X$ follows. $\qquad\square$

Lemma 4.8. *Let V be a finite set and $(\mathcal{F}, D[\mathcal{F}])$ a non-negative symmetric quadratic form with $D[\mathcal{F}] \subseteq l(V)$. Then, $(\mathcal{F}, D[\mathcal{F}])$ is a resistance form on V if and only if $D[\mathcal{F}] = l(V)$ and the two properties (RF-1) and (RF-5) are satisfied.*

Proof. First, suppose that $(\mathcal{F}, D[\mathcal{F}])$ is a resistance form. Then, (RF-1) and (RF-5) hold by definition. For any $f \in l(V)$, (RF-3) states the existence of $g \in D[\mathcal{F}] \subseteq l(V)$ such that $g\restriction_V = f$. Hence, $f \in D[\mathcal{F}]$ and $D[\mathcal{F}] = l(V)$ follows.

Now assume that $(\mathcal{F}, l(V))$ satisfies (RF-1) and (RF-5). For any $F \subseteq V$ and $f \in l(F)$, we have $\mathbf{1}_F \cdot f \in l(V)$ and (RF-3) follows.

Fix $x \in V$. Then, $\mathcal{F} + \delta_x^2$ is a non-negative symmetric quadratic form. If $(\mathcal{F} + \delta_x^2)(f) = 0$, it follows that $\mathcal{F}(f) = 0 = f(x)$. By (RF-1), f is constant and $f(x) = 0$ implies $f \equiv 0$. Hence, $\mathcal{F} + \delta_x^2$ is positive-definite, i.e. it is an inner product on $l(V)$. Since $l(V)$ is finite-dimensional, it is complete w.r.t. $\mathcal{F} + \delta_x^2$. Hence, (RF-2) follows.

We now show that (RF-5) holds. Let $x, y \in V$. If $x \neq y$, we have $|V| \geq 2$ and $\mathbf{1}_x$ is non-constant. Hence,

$$
0 < \frac{1}{\mathcal{F}(\mathbf{1}_x)} = \frac{|\mathbf{1}_x(x) - \mathbf{1}_x(y)|^2}{\mathcal{F}(\mathbf{1}_x)} \leq R_{\mathcal{F}}(x, y).
$$

By Lemma 4.7, there exist $x_1, \ldots, x_{n-1}$, such that for $x_0 := x, x_n := y$, $\mathcal{F}(\mathbf{1}_{x_k}, \mathbf{1}_{x_{k+1}}) < 0$ holds for all $k = 0, \ldots, n-1$. For $f \in l(X)$, f non-constant, the Cauchy-Schwarz inequality yields

$$(f(x) - f(y))^2 = \left(\sum_{k=1}^{n-1} f(x_k) - f(x_{k+1}) \right)^2$$

$$= \left(\sum_{k=1}^{n-1} (-\mathcal{F}(\mathbf{1}_{x_k}, \mathbf{1}_{x_{k+1}})) \cdot (f(x_k) - f(x_{k+1})) \cdot \frac{1}{(-\mathcal{F}(\mathbf{1}_{x_k}, \mathbf{1}_{x_{k+1}}))} \right)^2$$

$$\leq \left(\sum_{k=1}^{n-1} (-\mathcal{F}(\mathbf{1}_{x_k}, \mathbf{1}_{x_{k+1}})) \cdot (f(x_k) - f(x_{k+1}))^2 \right) \cdot \left(\sum_{k=1}^{n-1} \frac{1}{(-\mathcal{F}(\mathbf{1}_{x_k}, \mathbf{1}_{x_{k+1}}))} \right)$$

$$\leq \mathcal{F}(f) \cdot \left(\sum_{k=1}^{n-1} \frac{1}{(-\mathcal{F}(\mathbf{1}_{x_k}, \mathbf{1}_{x_{k+1}}))} \right)$$

and

$$R_{\mathcal{F}}(x, y) \leq \sum_{k=1}^{n-1} \frac{1}{(-\mathcal{F}(\mathbf{1}_{x_k}, \mathbf{1}_{x_{k+1}}))} < \infty$$

follows. $\qquad \square$

Lemma 4.9. *Let V be a finite set. A quadratic form $\mathcal{F} : l(V) \to \mathbb{R}$ is a resistance form on V if and only if there exists a connected graph $G = (V, c)$ such that $\mathcal{F}$ is its energy form. In particular, for $x \neq y$, we have*

$$c(x, y) = -\mathcal{F}(\mathbf{1}_x, \mathbf{1}_y) \tag{4.4}$$

and $c_x = \mathcal{F}(\mathbf{1}_x)$. Furthermore,

$$R_{\mathcal{F}}(x, y) = R_G(x, y) \tag{4.5}$$

holds for all $x, y \in V$.

Proof. Let $G = (V, c)$ be a connected graph with resistance form $\mathcal{E}$. Then, $\mathcal{E}$ is a non-negative, symmetric quadratic form and, by Lemma 1.13, satisfies (RF-1). For $x, y \in V$, $f \in l(V)$, we have
$$|m(f)(x) - m(f)(y)| \leq |f(x) - f(y)|$$
and thus $\mathcal{E}(m(f)) \leq \mathcal{E}(f)$.

If $\mathcal{F}$ is a resistance form on V, let $c(x, x) := 0$ for all $x \in V$. Lemma 4.3 states that, for $x, y \in V, x \neq y$,
$$c(x, y) := -\mathcal{F}(\mathbf{1}_x, \mathbf{1}_y)$$
is non-negative and $c(x, y) = c(y, x)$ follows by the symmetry of $\mathcal{F}$. Hence, the graph $G = (V, c)$ is well-defined and we have

$$\mathcal{F}(\mathbf{1}_x) = \mathcal{F}(\mathbf{1}_x, 1) - \sum_{y \neq x} \mathcal{F}(\mathbf{1}_x, \mathbf{1}_y) = 0 - \sum_{y \neq x} (-c(x, y)) = \sum_{y \in V} c(x, y) = c_x.$$

Lemma 4.7 implies that G is connected. Furthermore, the energy form $\mathcal{E}$ of G is given by

$$\mathcal{E}(f) = \frac{1}{2} \sum_{x,y \in V} c(x,y) \cdot (f(x) - f(y))^2 = \frac{1}{2} \sum_{x,y \in V} (-\mathcal{F}(\mathbf{1}_x, \mathbf{1}_y)) \cdot (f(x) - f(y))^2 = \mathcal{F}(f)$$

for every $f \in l(V)$, where the last equality is also due to Lemma 4.7.

Corollary 2.16 states that the effective resistance of G is given by

$$R_G(x,y) = \max \left\{ \frac{(f(x) - f(y))^2}{\mathcal{E}(f)} \mid f \in l(V), \mathcal{E}(f) > 0 \right\} = R_{\mathcal{F}}(x,y).$$

$\square$

Corollary 4.10 ([23, Theorem 2.1.12]). *Let $\mathcal{F}_1, \mathcal{F}_2$ be resistance forms on a finite set V. Then, $R_{\mathcal{F}_1} = R_{\mathcal{F}_2}$ if and only if $\mathcal{F}_1 = \mathcal{F}_2$.*

Proof. Let $G_1 = (V, c_1)$ and $G_2 = (V, c_2)$ be graphs with energy forms $\mathcal{F}_1$ and $\mathcal{F}_2$, respectively. By Lemma 4.7, we have $\mathcal{F}_1 = \mathcal{F}_2$ if and only if $c_1(x,y) = -\mathcal{F}_1(\mathbf{1}_x, \mathbf{1}_y) = -\mathcal{F}_2(\mathbf{1}_x, \mathbf{1}_y) = c_2(x,y)$ for all $x, y \in V, x \neq y$. This is equivalent to $G_1 = G_2$ which in turn is equivalent to $R_{\mathcal{F}_1} = R_1 = R_2 = R_{\mathcal{F}_2}$ by Remark 2.36. $\square$

4.2 Minimizers and Potentials

Proposition 4.11 ([23, Lemma 2.2.2]). *Let $(\mathcal{F}, D[\mathcal{F}])$ be a resistance form on a set X and $F \subseteq X$ a nonempty finite subset. Then, for every $f \in l(F)$, there exists a unique element $h_F(f) \in D[\mathcal{F}]$ such that $h_F(f){\upharpoonright}_F = f$ and*

$$\mathcal{T}(h_F(f)) = \min \{\mathcal{F}(g) \mid g \in D[\mathcal{F}], g{\upharpoonright}_F = f\}. \tag{4.6}$$

$h_F(f)$ is uniquely characterized by $h_F(f){\upharpoonright}_F = f$ and

$$\mathcal{F}(h_F(f), h) = 0 \ \forall \, h \in D[\mathcal{F}], h{\upharpoonright}_F \equiv 0.$$

Furthermore, $h_F : l(F) \to D[\mathcal{F}]$ is linear.

Proof. Fix $y \in F$ and let

$$A_f := \{g \in D[\mathcal{F}] \mid g{\upharpoonright}_F = f\}.$$

By (RF-3), $A_f \neq \emptyset$ and (RF-2) implies that $H_y = (D[\mathcal{F}], \mathcal{F} + \delta_y^2)$ is a Hilbert space with norm $\|g\| = \sqrt{\mathcal{F}(g) + g(y)^2}$. By Lemma 4.4, convergence in H_y implies point-wise convergence. Hence, A_f is closed in H_y. Since A_f is also convex, Theorem 1.19 implies the existence of a unique element $h_F(f) \in A_f$ which minimizes $\|\cdot\|$. Since $g(y) = f(y)$ for all $g \in A_f$, we have

$$\mathcal{F}(h_F(f)) + f(y)^2 = \|h_F(f)\|^2 = \min_{g \in A_f} \|g\|^2 = \min_{g \in A_f} \left(\mathcal{F}(g) + f(y)^2\right) = f(y)^2 + \min_{g \in A_f} \mathcal{F}(g)$$

and the first claim follows.

Let $h \in D[\mathcal{F}]$ such that $h\restriction_F \equiv 0$ and $d \in \mathbb{R}$, $d \neq 0$. Then, $h_F(f) + d \cdot h \in A_f$ and we have

$$\mathcal{F}(h_F(f)) \leq \mathcal{F}(h_F(f) + d \cdot h) = \mathcal{F}(h_F(f)) + 2d \cdot \mathcal{F}(h_F(f), h) + d^2 \mathcal{F}(h).$$

Hence,

$$-2 \cdot \mathcal{F}(h_F(f), h) \leq d \cdot \mathcal{F}(h)$$

for all $d \in \mathbb{R}$, $d \neq 0$ and it follows that $\mathcal{F}(h_F(f), h) = 0$.

Conversely, let $g \in A_f$ such that $\mathcal{F}(g, h) = 0$ for all $h \in D[\mathcal{F}]$ such that $h\restriction_F \equiv 0$. For any $g' \in A_f$, we have $g' - g\restriction_F = 0$ and thus

$$\mathcal{F}(g') = \mathcal{F}(g + (g' - g)) = \mathcal{F}(g) + 2\mathcal{F}(g, g' - g) + \mathcal{F}(g' - g)$$
$$= \mathcal{F}(g) + \mathcal{F}(g' - g) \geq \mathcal{F}(g).$$

It follows that $g = h_F(f)$ by uniqueness.

Now we show that h_F is linear. Let $a, b \in \mathbb{R}$, $f, g \in l(F)$. Then, $a \cdot h_F(f) + b \cdot h_F(g) \in A_{af+bg}$ and, for $h \in D[\mathcal{F}]$ such that $h\restriction_F \equiv 0$, we have

$$\mathcal{F}(a \cdot h_F(f) + b \cdot h_F(g), h) = a^2 \mathcal{F}(h_F(f), h) + b^2 \mathcal{F}(h_F(g), h) = 0.$$

Hence, $a \cdot h_F(f) + b \cdot h_F(g) = h_F(af + bg)$. $\qquad\qquad\qquad\qquad\qquad\qquad\square$

The statement of Proposition 4.11 is essential to the theory of resistance forms. It allows us to verify the existence of minimizers and potentials analogous to the case of effective resistances on finite graphs (cf. Proposition 2.14). It will also allow us to define the trace of a resistance form on a finite set.

Corollary 4.12. *For $x, y \in X$, $x \neq y$, there exists a unique element $\psi_{D[\mathcal{F}]}^{xy} \in D[\mathcal{F}]$ such that $\psi_{D[\mathcal{F}]}^{xy}(x) = 1$, $\psi_{D[\mathcal{F}]}^{xy}(y) = 0$ and*

$$\mathcal{F}\left(\psi_{D[\mathcal{F}]}^{xy}\right) = \min\left\{\mathcal{F}(f) \mid f \in D[\mathcal{F}], f(x) = 1, f(y) = 0\right\}. \tag{4.7}$$

In particular, the supremum in (RF-4) is a maximum and is attained by

$$R_{(\mathcal{F}, D[\mathcal{F}])}(x, y) = \mathcal{F}\left(\psi_{D[\mathcal{F}]}^{xy}\right)^{-1}. \tag{4.8}$$

We call $\psi_{D[\mathcal{F}]}^{xy}$ **minimizer** *of $(\mathcal{F}, D[\mathcal{F}])$.*

Proof. The first claim follows directly form Proposition 4.11 by choosing $F = \{x, y\}$ and $f = \mathbf{1}_x$.

The fact that (4.8) holds follows by the same argument as in Corollary 2.16 but we include it at this point for the convenience of the reader. For $f \in D[\mathcal{F}]$ such that $f(x) \neq f(y)$, let

$$\tilde{f}(v) := \frac{f(v) - f(y)}{f(x) - f(y)}, \quad v \in x.$$

Then, $\widetilde{f}(x) = 1$, $\widetilde{f}(y) = 0$ and

$$\frac{(f(x) - f(y))^2}{\mathcal{F}(f)} = \frac{1}{\mathcal{F}(\widetilde{f})} \le \frac{1}{\mathcal{F}\left(\psi^{xy}_{D[\mathcal{F}]}\right)} = \frac{(\psi^{xy}_{D[\mathcal{F}]}(x) - \psi^{xy}_{D[\mathcal{F}]}(y))^2}{\mathcal{F}\left(\psi^{xy}_{D[\mathcal{F}]}\right)}$$

follows. This implies (4.8). $\qquad\square$

Corollary 4.13. *For $x, y \in X$, $x \ne y$, there exists $\phi^{xy}_{D[\mathcal{F}]} \in D[\mathcal{F}]$ such that $\phi^{xy}_{D[\mathcal{F}]}(y) = 0$* *and*

$$R_{(\mathcal{F},D[\mathcal{F}])}(x, y) = \mathcal{F}\left(\phi^{xy}_{D[\mathcal{F}]}\right) = \phi^{xy}_{D[\mathcal{F}]}(x). \tag{4.9}$$

*We call $\phi^{xy}_{D[\mathcal{F}]}$ **potential** of $(\mathcal{F}, D[\mathcal{F}])$.*

Proof. For convenience, write ϕ and ψ instead of $\phi^{xy}_{D[\mathcal{F}]}$ and $\psi^{xy}_{D[\mathcal{F}]}$, respectively. Let $\phi := \psi / \mathcal{F}(\psi)$. Then, $\phi(y) = 0$,

$$\phi(x) = \mathcal{F}(\psi)^{-1} = R_{\mathcal{F}}(x, y)$$

and

$$\mathcal{F}(\phi) = \frac{\mathcal{F}(\psi)}{\mathcal{F}(\psi)^2} = \mathcal{F}(\psi)^{-1} = R_{\mathcal{F}}(x, y).$$

$\qquad\square$

If $\mathcal{F}$ is a resistance form on a finite set, it is the energy form of a finite graph G. In this case $\phi^{xy}_{D[\mathcal{F}]}$ is exactly the unit potential of G and is its corresponding minimizer, see Proposition 2.14.

Definition 4.14. For non-constant $f \in D[\mathcal{F}]$, let

$$S^{xy}(f) := \frac{|f(x) - f(y)|^2}{\mathcal{F}(f)}. \tag{4.10}$$

Lemma 4.15. *For $f \in D[\mathcal{F}]$, we have*

$$S^{xy}(f) = \sup_{\substack{g \in D[\mathcal{F}] \\ \mathcal{F}(g) > 0}} S^{xy}(g), \tag{4.11}$$

if and only if there exist $a, b \in \mathbb{R}$, $a \ne 0$ such that

$$f = \psi^{xy}_{D[\mathcal{F}]} \cdot a + b.$$

In particular, $\phi^{xy}_{D[\mathcal{F}]}$ is the unique maximizer of S^{xy} in $D[\mathcal{F}]$ with boundary values $\phi^{xy}_{D[\mathcal{F}]}(x) = R_{\mathcal{F}}(x, y)$ and $\phi^{xy}_{D[\mathcal{F}]}(y) = 0$.

Proof. Let $f \in D[\mathcal{F}]$ such that $\mathcal{F}(f) > 0$. W.l.o.g. assume that $f(x) \neq f(y)$, since otherwise $S^{xy}(f) = 0$. For

$$\tilde{f}(z) := \frac{f(z) - f(y)}{f(x) - f(y)},$$

we have $\tilde{f}(x) = 1$ and $\tilde{f}(y) = 0$. It follows that

$$S^{xy}(f) = \frac{(f(x) - f(y))^2}{\mathcal{F}(f)} = \frac{(f(x) - f(y))^2}{\mathcal{F}(f - f(y))} = \frac{1}{\mathcal{F}(\tilde{f})} \leq \frac{1}{\mathcal{F}(\psi^{xy}_{D[\mathcal{F}]})} \, .$$

By uniqueness of $\psi^{xy}_{D[\mathcal{F}]}$, we have equality if and only if $\tilde{f} = \psi^{xy}_{D[\mathcal{F}]}$, i.e.

$$f(z) = \psi^{xy}_{D[\mathcal{F}]} \cdot (f(x) - f(y)) + f(y).$$

Here, $a := f(x) - f(y)$ and $b := f(y)$ can be chosen arbitrarily as long as $a \neq 0$. The uniqueness of $\phi^{xy}_{D[\mathcal{F}]}$ follows as well. $\qquad\square$

Proposition 4.16. *For $x, y \in V$, we have*

$$\mathcal{F}(\phi^{xy}_{D[\mathcal{F}]}, f) = f(x) - f(y) \tag{4.12}$$

for all $f \in D[\mathcal{F}]$, i.e. $\phi^{xy}_{D[\mathcal{F}]} = g^{xy}_y$ from Proposition 4.5.

Proof. Fix $x, y \in V$ and let $g = g^{xy}_y$ for simplicity. Then, $g(y) = 0$ by Proposition 4.5. Note that g is not constant if $x \neq y$. Otherwise, we have $0 = \mathcal{F}(f, g) = f(x) - f(y)$ for all $f \in D[F]$ which is impossible due to (RF-3). Hence,

$$g(x) = g(x) - g(y) = \mathcal{F}(g, g) > 0.$$

Let $f \in D[\mathcal{F}]$. If $f(x) = f(y)$, then $S^{xy}(f) = 0 < S^{xy}(g)$. If $f(x) \neq f(y)$, define $\tilde{g} := g/g(x)$ and $\tilde{f} := f/(f(x) - f(y))$. Then, $\tilde{g}(x) - \tilde{g}(y) = \tilde{f}(x) - \tilde{f}(y) = 1$ and

$$\begin{aligned}
S^{xy}(f)^{-1} &= \mathcal{F}(\tilde{f}) = \mathcal{F}(\tilde{g}) + 2\mathcal{F}(\tilde{g}, \tilde{f} - \tilde{g}) + \mathcal{F}(\tilde{f} - \tilde{g}) \\
&= \mathcal{F}(\tilde{g}) + \frac{2}{g(x)} \cdot \underbrace{\left(\left(\tilde{f}(x) - \tilde{g}(x)\right) - \left(\tilde{f}(y) - \tilde{g}(y)\right) \right)}_{=0} + \mathcal{F}(\tilde{f} - \tilde{g}) \\
&\geq \mathcal{F}(\tilde{g}) = S^{xy}(g)^{-1}.
\end{aligned}$$

Hence,

$$S^{xy}(g) = \max_{\substack{f \in D[\mathcal{F}] \\ \mathcal{F}(f) > 0}} S^{xy}(f) = S^{xy}(\phi^{xy}_{D[\mathcal{F}]}).$$

Furthermore,

$$g(x) = \frac{|g(x) - g(y)|^2}{\mathcal{F}(g)} = S^{xy}(g) = S^{xy}(\phi^{xy}_{D[\mathcal{F}]}) = R_{\mathcal{F}}(x, y)$$

which implies $\phi^{xy}_{D[\mathcal{F}]} = g$ by uniqueness of $\phi^{xy}_{D[\mathcal{F}]}$ and the claim follows. $\qquad\square$

4.3 Traces and Reproducing Sequences

We will now see that every resistance form on a countable set can be obtained as a limit from resistance forms on increasing finite sets.

Definition 4.17. Let $(\mathcal{F}, D[\mathcal{F}])$ be a resistance form on a set X and $\emptyset \neq W \subseteq X$ be finite. The *trace* $\mathcal{F}^W : l(W) \to \mathbb{R}$ of $\mathcal{F}$ is defined by

$$\mathcal{F}^W(f) := \mathcal{F}(h_W(f)) = \min_{\substack{g \in D[\mathcal{F}] \\ g\restriction_W = f}} \mathcal{F}(g).$$

Proposition 4.18 ([20, Lemmas 1.7 and 1.9]). *For any finite $\emptyset \neq W \subseteq X$, $(\mathcal{F}^W, l(W))$ is a resistance form on W with*

$$R_{\mathcal{F}}\restriction_W = R_{\mathcal{F}^W}.$$

Proof. First, we show that $\mathcal{F}^W$ is a quadratic form. For $\alpha \in \mathbb{R}$, $f \in l(W)$, the linearity of h_W yields

$$\mathcal{F}^W(\alpha f) = \mathcal{F}(h_W(\alpha f)) = \mathcal{F}(\alpha h_W(f)) = \alpha^2 \cdot \mathcal{F}(h_W(f)) = \alpha^2 \cdot \mathcal{F}^W(f)$$

and, for $f, g \in l(W)$,

$$\begin{aligned}
\mathcal{F}^W(f+g) - \mathcal{F}^W(f-g) &= \mathcal{F}(h_W(f+g)) + \mathcal{F}(h_W(f-g)) \\
&= \mathcal{F}(h_W(f) + h_W(g)) + \mathcal{F}(h_W(f) - h_W(g)) \\
&= 2\mathcal{F}(h_W(f)) + 2\mathcal{F}(h_W(g)) = 2\mathcal{F}^W(f) + 2\mathcal{F}^W(g).
\end{aligned}$$

By Lemma 4.8, $(\mathcal{F}^W, l(W))$ is a resistance form on W if (RF-1) and (RF-5) hold. Clearly, $\mathbf{1}_W \in l(W)$ and $\mathcal{F}^W(f) = 0$ for constant f hold. Suppose that $\mathcal{F}^W(f) = 0$. Then, $\mathcal{F}(h_W(f)) = 0$ which implies that $h_W(f)$ is constant. Hence, $f = h_W(f)\restriction_W$ is constant. For any $f \in l(W)$, we have $m(f) \in l(W)$ and $m(h_W(f))\restriction_W = m(f)$. Hence,

$$\mathcal{F}^W(m(f)) = \min_{\substack{g \in D[\mathcal{F}] \\ g\restriction_W = m(f)}} \mathcal{F}(g) \leq \mathcal{F}(m(h_W(f))) \leq \mathcal{F}(h_W(f)) = \mathcal{F}^W(f)$$

and it follows that $\mathcal{F}^W$ is a resistance form on W.

Let $x, y \in W$, $x \neq y$. Denote by ψ and ψ_W the unique minimizers of $\mathcal{F}$ and $\mathcal{F}^W$ such that $R_{\mathcal{F}}(x, y) = \mathcal{F}(\psi)^{-1}$ and $R_{\mathcal{F}^W}(x, y) = \mathcal{F}^W(\psi_W)^{-1}$. Then,

$$\mathcal{F}^W(\psi_W) = \min_{\substack{f \in l(W) \\ f(x)=1, f(y)=0}} \mathcal{F}^W(f) \leq \mathcal{F}^W(\psi\restriction_W) = \min_{\substack{f \in D[\mathcal{F}] \\ f\restriction_W = \psi}} \mathcal{F}(f) \leq \mathcal{F}(\psi)$$

and

$$\mathcal{F}(\psi) = \min_{\substack{f \in D[\mathcal{F}] \\ f(x)=1, f(y)=0}} \mathcal{F}(f) \leq \mathcal{F}(h_W(\psi_W)) = \mathcal{F}^W(\psi_W).$$

Hence, $\mathcal{F}(\psi) = \mathcal{F}^W(\psi_W)$ and the claim follows. In particular, $h_W(\psi_W) = \psi$. $\qquad\square$

Remark 4.19. Let $\mathcal{F}$ be a resistance form on a finite set V. Furthermore, let (V, c) be the graph with energy form $\mathcal{F}$ and $W := V \setminus \{x^*\}$. Then, the trace $\mathcal{F}^W$ of $\mathcal{F}$ is exactly the energy form of (W, c') as defined by the Star-Mesh transform (Theorem 2.6), namely

$$\mathcal{F}^W(\mathbf{1}_x, \mathbf{1}_y) = \mathcal{F}(\mathbf{1}_x, \mathbf{1}_y) - \frac{\mathcal{F}(\mathbf{1}_x, \mathbf{1}_{x^*}) \cdot \mathcal{F}(\mathbf{1}_{x^*}, \mathbf{1}_y)}{\mathcal{F}(\mathbf{1}_{x^*})} \tag{4.13}$$

Definition 4.20. We say a sequence of resistance forms $(\mathcal{F}_n)_{n \in \mathbb{N}}$ on non-empty finite sets V_n is a ***reproducing sequence*** if $V_n \subseteq V_{n+1}$ and $R_{\mathcal{F}_{n+1}} \restriction_{V_n} = R_{\mathcal{F}_n}$.

For any resistance form $(\mathcal{F}, D[\mathcal{F}])$ on a countable set V and any finite exhaustion $(V_n)_{n \in \mathbb{N}}$ of V, Proposition 4.18 implies that $(\mathcal{F}^{V_n})_{n \in \mathbb{N}}$ is a reproducing sequence.

Proposition 4.21. *Let $(\mathcal{F}_n)_{n \in \mathbb{N}}$ be a reproducing sequence on sets V_n. Define $V = \bigcup_{n \in \mathbb{N}} V_n$,*

$$D[\mathcal{F}] := \left\{ f \in l(V) \mid \lim_{n \to \infty} \mathcal{F}_n(f \restriction_{V_n}) < \infty \right\} \tag{4.14}$$

and, for $f \in D[\mathcal{F}]$,

$$\mathcal{F}(f) = \lim_{n \to \infty} \mathcal{F}_n(f \restriction_{V_n}). \tag{4.15}$$

Then, $(\mathcal{F}, D[\mathcal{F}])$ is a resistance form on V with

$$R_{\mathcal{F}}(x, y) = R_{\mathcal{F}_n}(x, y) \tag{4.16}$$

for all $x, y \in V_n$, $n \in \mathbb{N}$. We call $\mathcal{F}$ the **limit form** *of $(\mathcal{F}_n)_{n \in \mathbb{N}}$.*

Proof. Since the proof is mostly of technical nature, we refer the interested reader to Lemmas 2.2.2., 2.2.5. and Theorem 2.2.6. in [23]. $\qquad\square$

Remark 4.22. For every $f \in l(V)$, the sequence $(\mathcal{F}_n(f \restriction_{V_n}))_{n \in \mathbb{N}}$ is monotonically increasing. Indeed, for $n \in \mathbb{N}$, we have $h_{V_{n+1}}(f \restriction_{V_{n+1}}) \restriction_{V_n} = f \restriction_{V_n}$ and thus

$$\mathcal{F}_n(f \restriction_{V_n}) = \min_{\substack{g \in l(V) \\ g \restriction_{V_n} = f \restriction_{V_n}}} \mathcal{F}(g) \leq \mathcal{F}(h_{V_{n+1}}(f \restriction_{V_{n+1}})) = \mathcal{F}_{n+1}(f \restriction_{V_{n+1}}).$$

Hence, $\lim_{n \to \infty} \mathcal{F}_n(f \restriction_{V_n})$ exists or is $+\infty$.

Lemma 4.23 ([23, Lemma 2.3.8]). *Let $(\mathcal{F}, D[\mathcal{F}])$ be a resistance form on a countable set V and $(V_n)_{n \in \mathbb{N}}$ a finite exhaustion. Then, the limit form of $(\mathcal{F}^{V_n})_{n \in \mathbb{N}}$ is $(\mathcal{F}, D[\mathcal{F}])$.*

Proof. Denote by $(\mathcal{F}^*, D[\mathcal{F}^*])$ the limit form of $(\mathcal{F}^{V_n})_{n \in \mathbb{N}}$.

For $f \in D[\mathcal{F}]$, we have

$$\lim_{n \to \infty} \mathcal{F}^{V_n}(f \restriction_{V_n}) = \lim_{n \to \infty} \min_{\substack{g \in D[\mathcal{F}] \\ g \restriction_{V_n}}} \mathcal{F}(g) \leq \mathcal{F}(f) < \infty$$

and $f \in D[\mathcal{F}^*]$ follows. In particular, $\mathcal{F}^*(f) \leq \mathcal{F}(f)$.

For $f \in D[\mathcal{F}^*]$, let $f_n := h_{V_n}(f \restriction_{V_n}) \in D[\mathcal{F}]$. It follows that $f_n \to f$ point-wise and $\lim_{n \to \infty} \mathcal{F}(f_n) < \infty$. Lemma 4.6 implies $f \in D[\mathcal{F}]$ and

$$\mathcal{F}(f) \leq \liminf_{n \to \infty} \mathcal{F}(f_n) = \liminf_{n \to \infty} \mathcal{F}^{V_n}(f \restriction_{V_n}) = \mathcal{F}^*(f).$$

$\qquad\square$

In particular, Lemma 4.23 implies that two resistance forms on the same set coincide if and only if they have the same traces.

Corollary 4.24. *Let $(\mathcal{F}_1, D_1)$, $(\mathcal{F}_2, D_2)$ be two resistance forms on the same set X. Then, the following are equivalent:*

(i) $(\mathcal{F}_1, D_1) = (\mathcal{F}_2, D_2)$

(ii) $(\mathcal{F}_1)^W = (\mathcal{F}_2)^W$ for all finite $W \subseteq X$

(iii) $R_{\mathcal{F}_1} = R_{\mathcal{F}_2}$

Proof. Clearly, $(i) \Rightarrow (ii)$ and $(i) \Rightarrow (iii)$ hold. By Lemma 4.23, $(ii) \Rightarrow (i)$ is also true. Lastly, $(iii) \Rightarrow (ii)$ follows from Corollary 4.10. $\qquad\square$

4.4 Resistance Metrics

Definition 4.25. Let X be any set. A function $R : X \times X \to \mathbb{R}_{\geq 0}$ is called **resistance metric** on X if, for any finite $W \subseteq X$, there exists a resistance form $\mathcal{F}_W$ on W such that

$$R\!\restriction_W = R_{\mathcal{F}_W}.$$

By Lemma 4.8, R is a resistance metric on X if and only if, for any $W \subseteq X$, there exists a connected graph $G_W = (W, c_W)$ such that $R\!\restriction_W = R_{G_W}$. In other words, R is a resistance metric on X if and only if every finite subspace of (X, R) is an effective resistance space (ERS). This way we could have defined resistance metrics without using the theory of resistance forms.

Remark 4.26. A resistance metric is a metric on X since every restriction to a finite subset of X is a metric.

Suppose that X is a finite set. If R is a resistance metric on X, then (X, R) is by definition an ERS. Conversely, if (X, R) is an ERS, then due to the Star-Mesh transform, every metric subspace is also an ERS which means that R is a resistance metric. Hence, the notion of resistance metrics is an extension of ERS to arbitrary sets. However, even on countable sets, the theory of resistance metrics is much more general than the classic notion of effective resistances and may not be directly related to graphs/networks. We start our investigation of resistance metrics by considering some examples.

Example 4.27. The metric on $\mathbb{R}$ induced by the absolute value is a resistance metric. Indeed, for any finite $A \subseteq \mathbb{R}$ choose an ordering $A = \{a_1, \dots, a_n\}$ such that $a_1 < a_2 < \dots < a_n$. Then, $d\!\restriction_A$ is the effective resistance of the line graph shown in Figure 4.1.

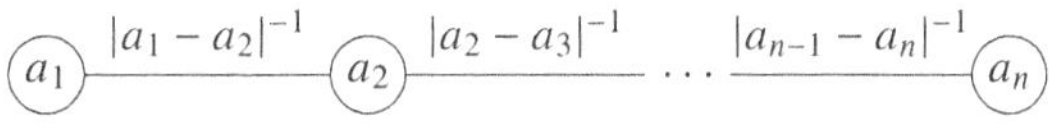

$$a_1 \; \overset{|a_1 - a_2|^{-1}}{\longrightarrow} \; a_2 \; \overset{|a_2 - a_3|^{-1}}{\longrightarrow} \; \dots \; \overset{|a_{n-1} - a_n|^{-1}}{\longrightarrow} \; a_n$$

Figure 4.1: The graph with effective resistance $R(a_i, a_j) = |a_i - a_j|$.

Example 4.28. The Euclidean metric on $\mathbb{R}^n$, $n > 1$, i.e.

$$d_E(x, y) = \sqrt{\sum_{k=1}^{n} (x_k - y_k)^2},$$

is no resistance metric on $\mathbb{R}^n$. Indeed, denote by e_k the k-th unit vector, i.e. $(e_k)_i = \mathbf{1}_k(i)$, and consider the set $A = \{0, e_1, e_2, e_1 + e_2\}$. Then, $d_E \restriction_A$ has the geodesic graph depicted in Figure 4.2. Example 2.60 implies that $d_E \restriction_A$ is no effective resistance since $\sqrt{2} \in (4/3, 2)$.

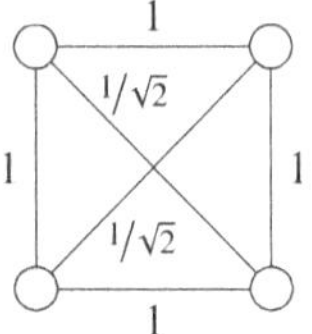

Figure 4.2: The geodesic graph of the 2-dimensional unit cube.

Example 4.29. Consider the discrete metric d_{disc} on $\mathbb{N}$, i.e. $d_{\mathrm{disc}}(x, y) = 1$ for all $x \neq y$. For any finite subset $A \subseteq \mathbb{N}$, $(A, d_{\mathrm{disc}} \restriction_A)$ is an ERS with associated graph (A, c_A) where

$$c_A(x, y) = \frac{2}{|A|} \cdot (1 - \mathbf{1}_x(y)).$$

cf. Example 2.5. Hence, d_{disc} is a resistance metric on $\mathbb{N}$.

Theorem 4.30 ([23, Theorems 2.3.4 and 2.3.6]). *Let V be a countable set. Then, $R : V \times V \to \mathbb{R}$ is a resistance metric on V if and only if there exists a resistance form $(\mathcal{F}, D[\mathcal{F}])$ on V such that $R = R_{\mathcal{F}}$.*

Proof. Suppose that R is a resistance metric on V and let $(V_n)_{n \in \mathbb{N}}$ be a finite exhaustion of V. Then, for every $n \in \mathbb{N}$, there exists a resistance forms $\mathcal{F}_n$ on V_n such that $R_{\mathcal{F}_n} = R \restriction_{V_n}$. Hence, $(\mathcal{F}_n)_{n \in \mathbb{N}}$ is a reproducing sequence with a limit form $(\mathcal{F}, D[\mathcal{F}])$. For $x, y \in V_n$, we then have $R(x, y) = R_{\mathcal{F}_n}(x, y) = R_{\mathcal{F}}(x, y)$.

Now assume that $(\mathcal{F}, D[\mathcal{F}])$ is a resistance form satisfying $R = R_{\mathcal{F}}$ and let $W \subseteq V$ be finite. Then, the trace $\mathcal{F}^W$ is a resistance form on W such that $R_{\mathcal{F}^W} = R \restriction_W$. Hence, R is a resistance metric. $\square$

Remark 4.31. The proof for the 'if' part does not use the countability of V. Applying Corollary 4.33 and using that $D[\mathcal{F}] \subseteq C(V, \sqrt{R})$, one can extend the 'only if' statement to uncountable V such that (V, R) is separable.

In fact, Kigami mentions that the statement holds for arbitrary sets V (cf. [23, Theorems 2.3.4 and 2.3.6]).

In order to characterize resistance metrics algebraically, we can apply our characterization of effective resistance spaces (Theorem 2.44) to every finite subset of X. On countable sets it suffices to check the conditions of Theorem 2.44 only on a finite exhaustion.

Theorem 4.32. *Let V be a countably infinite set V and $(V_n)_{n \in \mathbb{N}}$ a finite exhaustion of V. Then, a function $R : V \times V \to \mathbb{R}_{\geq 0}$ is a resistance metric on V if and only if $R_n := R{\upharpoonright}_{V_n}$ satisfies the criterion of Theorem 2.44 for all $n \in \mathbb{N}$.*

Proof. Let R be a resistance metric and $n \in \mathbb{N}$. By definition, $R_n = R_{\mathcal{F}_n}$ holds for some resistance form $\mathcal{F}_n$ on V_n. By Lemma 4.9, there exists a finite graph $G_n = (V_n, c_n)$ such that $R_{G_n} = R_{\mathcal{F}_n} = R_n$. Hence, R_n is a finite ERS and thus satisfies the criterion of Theorem 2.44.

Now assume that $R_n = R{\upharpoonright}_{V_n}$ satisfies the criterion of Theorem 2.44 for all $n \in \mathbb{N}$. Hence, R_n is the effective resistance of a finite graph $G_n = (V_n, c_n)$ with energy form $\mathcal{E}_n$. Let $W \subseteq V$ be any finite subset of V. Since $\bigcup V_n = V$, there exists $n \in \mathbb{N}$ such that $W \subseteq V_n$. Then, the trace $(\mathcal{E}_n)^W$ of $\mathcal{E}_n$ is a resistance form on W and satisfies

$$R_{(\mathcal{E}_n)^W}(x, y) = R_{\mathcal{E}_n}(x, y) = R_n(x, y) = R(x, y)$$

for all $x, y \in W$. It follows that R is a resistance metric. $\qquad\square$

Corollary 4.33. *Let R be a resistance metric on a any set V and $(\overline{V}, \overline{R})$ the completion of (V, R). Then, $\overline{R}$ is a resistance metric on $\overline{V}$.*

Remark 4.34. We present a proof which does not directly use resistance forms but relies much more on Graph theory and arguments of effective resistances on finite graphs. Using resistance forms, the statement may be easier to prove by showing that for $R_{\mathcal{F}} = R$, all $f \in D[\mathcal{F}]$ are continuous with respect to the metric $\sqrt{R}$.

Proof of Corollary 4.33. Let $v_\infty \in \overline{V} \setminus V$, $W := V \cup \{v_\infty\}$ and $(v_n)_{n \in \mathbb{N}}$ a sequence such that $\overline{R}(v_n, v_\infty) \to 0$. We claim that $\overline{R}{\upharpoonright}_W$ is a resistance metric on W. By definition, we need to show that $(F, \overline{R}{\upharpoonright}_F)$ is an ERS for every finite $F \subseteq W$. This holds if $v_\infty \notin F$ since then $F \subseteq V$ and (V, R) is a resistance metric. Now, suppose $v_\infty \in F$ and define $F_n := (F \setminus \{v_\infty\}) \cup \{v_n\}$ and $R_n := \overline{R}{\upharpoonright}_{F_n}$. Then, (F_n, R_n) is an ERS since $F_n \subseteq V$ and we denote by $G_n = (F_n, c_n)$ its associated graph. Furthermore, (F_n, R_n) satisfies the criterion of Theorem 2.44. For $n \in \mathbb{N}$ and $y \in F_n$, we define matrices $A_{y,n}, M_{y,n}, b_{y,n}$ similar to the context of Theorem 2.44 such that

$$A_{y,n} \cdot c_n(\cdot, y) = b_y \; \forall \, y \in F_n$$

and $\det(A_{y,n}) = \det(M_{y,n}) > 0$ for all $y \in F_n$. Since we have

$$K := \inf_{\substack{n \in \mathbb{N} \\ x, z \in F_n \\ x \neq z}} \overline{R}(x, z) > 0,$$

it follows by Proposition 2.34 that

$$0 < \det(M_{y,n})^{-1} = \sum_{\substack{T \text{ spanning} \\ \text{tree of } G_n}} \sqrt{\prod_{\substack{x,z \text{ adjacent} \\ \text{in } T}} c_n(x, z)} \leq \sum_{\substack{T \text{ spanning} \\ \text{tree of } G_n}} \sqrt{\prod_{\substack{x,z \text{ adjacent} \\ \text{in } T}} \frac{1}{\overline{R}(x, z)}}$$

$$\leq \sum_{\substack{T \text{ spanning} \\ \text{tree of } G_n}} \sqrt{\prod_{\substack{x,z \text{ adjacent} \\ \text{in } T}} K^{-1}} = \sum_{\substack{T \text{ spanning} \\ \text{tree of } G_n}} (\sqrt{K})^{1-|W|}$$

$$\leq \binom{|F|^2}{|F| - 1} (\sqrt{K})^{1-|F|} < \infty.$$

Let A_y, M_y, b_y be analogously defined for $(F, \overline{R}\restriction_F)$. We then have $\det(A_y) = \det(M_y) = \lim_{n\to\infty} \det(M_{y,n}) > 0$. Since c_n's columns consist of solutions of linear equation systems with coefficient matrices $A_{y,n} \to A_y$ entry-wise, it follows that

$$c(x, y) := \lim_{n\to\infty} \begin{cases} c_n(x, y) & , x \neq v_\infty \neq y \\ c_n(x, v_n) & , y = v_\infty \\ c_n(v_n, y) & , x = v_\infty \end{cases}.$$

exists and satisfies

$$A_y \cdot c(\cdot, y) = b_y \; \forall \, y \in F.$$

Since $c_n \geq 0$ entry-wise, we also have $c \geq 0$ entry-wise. Thus, $(F, \overline{R}\restriction_F)$ satisfies the criterion of Theorem 2.44 and is therefore an ERS. It follows that $\overline{R}\restriction_W$ is a resistance metric on W.

The above argument only uses the fact that R is a resistance metric on V and that v_∞ is a limit point of R. Hence, we can inductively extend the argument to $W = V \cup \{v_\infty^1, \ldots, v_\infty^n\}$ for $v_\infty^i \in \overline{V} \setminus V$. This already suffices for $\overline{R}$ to be a resistance metric on $\overline{V}$. Indeed, take any finite $F \subseteq \overline{V}$ and let $W := V \cup F$. By the above argument, $\overline{R}\restriction_W$ is a resistance metric on W. Hence, $(F, \overline{R}\restriction_F)$ is an ERS. The claim follows. $\qquad\square$

4.5 The Limit Graph of Resistance Metrics

Let V be a countably infinite set, R be a resistance metric on V and $(V_n)_{n\in\mathbb{N}}$ a finite exhaustion of V with associated graphs $G_n = (V_n, c_n)$ such that their effective resistance is $R_n = R\restriction_{V_n}$. We will show that there exists a **limit graph** G_R which is naturally associated to R.

Note that we do not require R to be associated to any kind of graph. Neither do we make any assumptions regarding the domain of the associated resistance form.

4.5.1 Existence and Definition

Proposition 4.35. *Let V_1, V_2 be two finite sets such that $V_1 \subseteq V_2$ and let (V_1, c_1) and (V_2, c_2) be two graphs with effective resistances R_1 and R_2. If $R_1(x, y) = R_2(x, y)$ for all $x, y \in V_1$, then*

$$c_1(x, y) \geq c_2(x, y) \; \forall \, x, y \in V_1$$

and

$$(c_1)_x \leq (c_2)_x \; \forall \, x \in V_1.$$

In particular,

$$\frac{c_1(x, y)}{(c_1)_x} \geq \frac{c_2(x, y)}{(c_2)_x} \; \forall \, x, y \in V_1$$

Proof. The claim follows from (2.8) and (2.9) by successively applying the Star-Mesh transform (Theorem 2.6) to each vertex in $V_2 \setminus V_1$. $\qquad\square$

Proposition 4.36. *For $x, y \in V$, the limit*

$$\lim_{n \to \infty} c_n(x, y) \in [0, +\infty)$$

exists and is independent of the choice of $(V_n)_{n \in \mathbb{N}}$. Furthermore,

$$\lim_{n \to \infty} (c_n)_x \in (0, +\infty]$$

exists for all $x \in V$ and is independent of the choice of $(V_n)_{n \in \mathbb{N}}$.

Proof. Let $x, y \in V_m$. By Corollary 4.35, we have

$$c_n(x, y) \geq c_{n+1}(x, y) \; \forall \, n \geq m$$

Since $c_n(x, y) \geq 0$ for all $n \geq m$, we see that $\lim_{n \to \infty} c_n(x, y)$ does indeed exist. Its independence regarding the choice of (V_n) follows by the same generic argument as seen in the proof of Lemma 3.2.

The second statement follows analogously since Corollary 4.35 also states that $(c_n)_x \leq (c_{n+1})_x$ for all $n \geq m$. The only difference is that we do not have an upper bound for the sequence and hence it may be possible that $(c_n)_x \to \infty$. $\qquad\qquad\square$

Definition 4.37. Using the notation of Proposition 4.36, we define the *limit graph* of R to be $G_R := (V, c)$ with

$$c(x, y) := \lim_{n \to \infty} c_n(x, y) \, , \; x, y \in V.$$

Remark 4.38. The definition of G_R does not depend on the choice of the exhaustion $(V_n)_{n \in \mathbb{N}}$.

Furthermore, G_R is a weighted graph as defined in Section 1.4. Indeed for $x, y \in V$, we have $c(x, x) = \lim_{n \to \infty} c_n(x, x) = 0$ and $c(x, y) = \lim_{n \to \infty} c_n(x, y) = \lim_{n \to \infty} c_n(y, x) = c(y, x)$ because all (V_n, c_n) are graphs.

It seems natural to pose two questions. Is R a resistance metric of G_R? Does it have any connection to the random walk on G_R? The following examples show that G_R may in some cases not allow for a well-defined notion the Laplacian. However, we will see at a later point (Proposition 4.61), that if R already "comes from" a graph G, then $G_R = G$.

Example 4.39. We continue Example 4.29 and consider the discrete metric d_{disc} on $V = \mathbb{N}$. For any finite exhaustion $(V_n)_{n \in \mathbb{N}}$, the graph $G_n = (V_n, c_n)$ is given by

$$c_n(x, y) = \frac{2}{|V_n|}(1 - \mathbf{1}_x(y)).$$

It follows that $c(x, y) = 0$ for all $x, y \in V$ since $|V_n| \to \infty$. Thus, $G_{d_{\text{disc}}}$ is completely disconnected and its Laplacian is not well-defined. Hence, it is not clear how to define the effective resistance of G.

Example 4.40. Consider $V = \mathbb{N}$ equipped with the star metric V, i.e.,

$$d_{\star}(x, y) = \begin{cases} 0 & , x = y \\ 1 & , x = 1 \text{ or } y = 1 \\ 2 & , x \neq 1 \neq y \end{cases}.$$

For the finite exhaustion (V_n) of V choose $V_n = \{1, \ldots, n\}$. Then, $d_{\star}\!\restriction_{V_n}$ is the effective resistance of (V_n, c_n) where

$$c_n(x, y) = \begin{cases} 1 & , x = 1, y \neq 1 \text{ or } x \neq 1, y = 1 \\ 0 & , \text{otherwise} \end{cases}.$$

This is due to the fact that (V_n, c_n) is a tree which implies that the effective resistance equals the geodesic metric, see Proposition 2.30. Hence, $G_{d_{\star}}$ is connected but since $c_1 = \sum_{n \in \mathbb{N}} c(1, n) = \infty$, its Laplacian is not well-defined.

Remark 4.41. The approach taken in this section should not be confused with considering free or wired exhaustions of a graph G. In fact, it is somewhat of a dual method.

Suppose we are investigating the free effective resistance R^F of a graph G. When considering a free exhaustion of G, we restrict its edge weights to a finite subset of vertices to get $G_n^F := G\!\restriction_{V_n}$. In general, this leads to a sequence of graphs with fixed edge weights and changing effective resistances which tend to R^F.

Using our current approach, we restrict the metric R^F to a finite subset of vertices. This yields graphs G_n with fixed effective resistance and changing edge weights. In general, G_n is not a subgraph of G.

4.5.2 Weak Convergence of Random Walks

For $x \in V$, denote by $\mathbf{P}_x^n$ the distributions of the random walk on G_n. We will now investigate under which conditions the sequence $(\mathbf{P}_x^n)_{n \in \mathbb{N}}$ has a weak limit point.

Prohorov's theorem [6, Theorem 8.6.2] states that on a separable and complete metric space Ω, a sequence $(\mu_n)_{n \in \mathbb{N}}$ of probability measures on the Borel-σ-field is sequentially compact with respect to weak convergence in the space of probability measures on Ω if and only if the sequence is tight, i.e.,

$$\forall \, \varepsilon > 0 \; \exists \, K_\varepsilon \subset\subset \Omega \; \forall \, n \in \mathbb{N} : \mu_n(\Omega \setminus K_\varepsilon) < \varepsilon. \tag{4.17}$$

Recall that we consider $\Omega = V^{\mathbb{N}_0}$ with the product topology induced by the discrete metric on V. Then, Ω is metrizable with metric ρ as in (1.39). We leave the verification that (Ω, ρ) is complete and separable as an exercise. Although $\mathbf{P}_x^n$ is originally defined on $(V_n)^{\mathbb{N}_0}$, we can interpret it as a Borel-measure on Ω since it is uniquely defined on all cylinder sets if we think of the vertices $V \setminus V_n$ as unreachable.

Since every subset of V is open, the product topology is generated by the set of all cylinder sets

$$C = \left\{ A_1 \times \ldots \times A_n \times V^{\mathbb{N}} \mid n \in \mathbb{N}, A_i \subseteq V \right\}.$$

Note the following properties of C.

$$\forall A, B \in C : A \cap B \in C \tag{4.18}$$

$$\forall A \in C \; \exists n \in \mathbb{N}, B_1, \ldots, B_n \in C : \Omega \setminus A = B_1 \cup \ldots \cup B_n \tag{4.19}$$

In particular, $\mathbf{1}_A$ is bounded and continuous for all $A \in C$.

We begin our investigation of when $(\mathbf{P}_x^n)_{n \in \mathbb{N}}$ is weakly convergent by observing a necessary condition.

Lemma 4.42. *Let $x \in V$. If $\lim_{n \to \infty} (c_n)_x = \infty$, then $\lim_{n \to \infty} \mathbf{P}_x^n[X_1 = y] = 0$ for all $y \in V$.*

Proof. Let $y \in V$, $y \neq x$ and $n \in \mathbb{N}$. By Corollary 2.20, we have $c_n(x, y) \leq R(x, y)^{-1}$ and thus

$$\mathbf{P}_x^n[X_1 = y] = \frac{c_n(x, y)}{(c_n)_x} \leq \frac{R(x, y)^{-1}}{(c_n)_x} .$$

It follows that $\lim_{n \to \infty} \mathbf{P}_x^n[X_1 = y] = 0$.

In the special case of $x = y$, we have $c_n(x, x) = 0$ and thus $\mathbf{P}_x^n[X_1 = x] = 0$ for all $n \in \mathbb{N}$. $\qquad \square$

Lemma 4.42 shows that $(\mathbf{P}_x^n)_{n \in \mathbb{N}}$ can not have a weak limit if $(c_n)_x \to \infty$. Indeed, suppose that $\mathbf{P}_x$ is a weak limit of $(\mathbf{P}_x^n)_{n \in \mathbb{N}}$. Since $\{X_1 = y\}, \{X_1 \in V\} \in C$, it follows that $\mathbf{1}_{\{X_1 = y\}}$ and $\mathbf{1}_{\{X_1 \in V\}}$ are bounded and continuous with respect to the topology of Ω. Since $\mu(A) = \int \mathbf{1}_A \, d\mu$ for any measure μ, we have

$$1 = \lim_{n \to \infty} \mathbf{P}_x^n[X_1 \in V] = \mathbf{P}_x[X_1 \in V] = \sum_{y \in V} \mathbf{P}_x[X_1 = y] = \sum_{y \in V} \lim_{n \to \infty} \mathbf{P}_x^n[X_1 = y] = 0$$

which is an obvious contradiction. It follows that the following condition (C) is necessary for the tightness of $(\mathbf{P}_x^n)_{n \in \mathbb{N}}$ and we will from now on assume that it is satisfied.

$$\lim_{n \to \infty} (c_n)_x < \infty \; \forall x \in V \tag{C}$$

Remark 4.43. Note that (C) implies that the metric space (V, R) must not contain limit points. Indeed, suppose that $R(x, y_n) \to 0$ for some sequence $(y_n)_{n \in \mathbb{N}}$ of vertices and $x \in V$. By the probabilistic representation (2.17) of R in (V_n, c_n), we have

$$0 \leq \frac{1}{(c_n)_x} \leq \frac{1}{(c_n)_x} \mathbf{E}_x^n \left[\sum_{k=0}^{\tau_{y_n} - 1} \mathbf{1}_x(X_k) \right] = R(x, y_n) \to 0$$

which implies $(c_n)_x \to \infty$.

Lemma 4.44. *The sequence $(\mathbf{P}_{x_0}^n)_{n \in \mathbb{N}}$ is tight if and only if*

$$\forall \varepsilon > 0 \; \exists m \in \mathbb{N} \; \forall n \in \mathbb{N} : \sum_{y \notin V_m} \frac{c_n(x, y)}{(c_n)_x} < \varepsilon \tag{T1}$$

holds for all $x \in V$.

Proof. First, assume that (T1) holds for all $x \in V$. Then, for $\delta > 0$ and $x \in V$, there exists $m = m(x) \in \mathbb{N}$ such that

$$\sup_{n \in \mathbb{N}} \sum_{y \notin V_m} \frac{c_n(x, y)}{(c_n)_x} < \delta \,.$$

We define $N(x, \delta) := V_m$ where m is the smallest number satisfying the above inequality. For a finite set $F \subseteq V$, we define

$$N(F, \delta) := \bigcup_{x \in F} N(x, \delta/|F|).$$

Then,

$$\sup_{n \in \mathbb{N}} \sum_{x \in F} \sum_{y \notin N(F, \delta)} \frac{c_n(x, y)}{(c_n)_x} \le \sum_{x \in F} \left(\sup_{n \in \mathbb{N}} \sum_{y \notin N(x, \delta)} \frac{c_n(x, y)}{(c_n)_x} \right) < \delta.$$

Now let $\varepsilon > 0$ and fix $x_0 \in V$. Furthermore, let $A_0 := \{x_0\}$ and for $n \in \mathbb{N}_0$

$$A_{n+1} = N(A_n, \varepsilon/2^{n+1}).$$

Finally, set $K_\varepsilon := \prod_{n \in \mathbb{N}_0} A_n$. Since all A_n are finite, K_ε is compact in Ω. Using the Law of total probability we compute

$$\begin{aligned}
\mathbf{P}_{x_0}^n(\Omega \setminus K_\varepsilon) &= \sum_{k=0}^{\infty} \mathbf{P}_{x_0}^n(A_0 \times \ldots \times A_{k-1} \times (V \setminus A_k) \times V^{\mathbb{N}}) \\
&= \sum_{k=1}^{\infty} \prod_{m=1}^{k-1} \mathbf{P}_{x_0}^n[X_m \in A_m \mid X_{m-1} \in A_{m-1}, \ldots, X_0 \in A_0] \\
&\qquad\qquad \cdot \mathbf{P}_{x_0}^n[X_k \notin A_k \mid X_{k-1} \in A_{k-1}, \ldots, X_0 \in A_0] \\
&\le \sum_{k=1}^{\infty} \mathbf{P}_{x_0}^n[X_k \notin A_k \mid X_{k-1} \in A_{k-1}, \ldots, X_0 \in A_0] \\
&\le \sum_{k=1}^{\infty} \sum_{x \in A_{k-1}} \mathbf{P}_{x_0}^n[X_k \notin A_k \mid X_{k-1} = x] \\
&= \sum_{k=1}^{\infty} \sum_{x \in A_{k-1}} \sum_{y \notin A_k} \frac{c_n(x, y)}{(c_n)_x} \\
&< \sum_{k=1}^{\infty} \frac{\varepsilon}{2^k} = \varepsilon
\end{aligned}$$

It follows that $(\mathbf{P}_{x_0}^n)_{n \in \mathbb{N}}$ is tight.

We use contraposition to prove that tightness implies (T1), i.e. assume that

$$\exists \, \varepsilon > 0 \, \forall \, m \in \mathbb{N} \, \exists \, n \in \mathbb{N} : \sum_{y \notin V_m} \frac{c_n(x, y)}{(c_n)_x} \ge \varepsilon$$

for some $x \in V$. Furthermore let $K \subset\subset \Omega$ be any compact set. Since $\Omega = V^{\mathbb{N}_0}$ is equipped with the product topology of the discrete topology on V, the projection $\pi_1 : \Omega \to V$,

$\omega \mapsto \omega_1$ is continuous. Hence, $\pi_1(K) \subseteq V$ is compact and thus finite. It follows that there exists $m \in \mathbb{N}$ such that $\pi_1(K) \subseteq V_m$. By our assumption above there exists $n \in \mathbb{N}$ such that

$$\sum_{y \notin V_m} \frac{c_n(x, y)}{(c_n)_x} \geq \varepsilon$$

. Hence,

$$\mathbf{P}^n_{x_0}(\Omega \setminus K) \geq \mathbf{P}^n_{x_0}[X_1 \notin \pi_1(K)] \geq \mathbf{P}^n_{x_0}[X_1 \notin V_m]$$
$$= \sum_{y \notin V_m} \mathbf{P}^n_{x_0}[X_1 = y] = \sum_{y \notin V_m} \frac{c_n(x, y)}{(c_n)_x} \geq \varepsilon.$$

This implies that the sequence $(\mathbf{P}^n_{x_0})_{n \in \mathbb{N}}$ is not tight. $\qquad\square$

Lemma 4.45. *Let $x \in V$. Under the assumption (C), the condition (T1) is equivalent to*

$$\lim_{n \to \infty} (c_n)_x = c_x, \tag{T2}$$

where $c_x = \sum_{y \in V} c(x, y)$ and $c(x, y) = \lim_{n \to \infty} c_n(x, y)$.

Proof. Let $\varepsilon > 0$ and assume that (T2) holds. Then,

$$\lim_{n \to \infty} \frac{c_n(x, y)}{(c_n)_x} = \frac{c(x, y)}{c_x}.$$

and there exists an $m \in \mathbb{N}$ such that

$$1 - \sum_{y \in V_m} \frac{c(x, y)}{c_x} < \varepsilon.$$

Let $n \in \mathbb{N}$. For $n \leq m$ we have $V_n \subseteq V_m$ and thus

$$\sum_{y \notin V_m} \frac{c_n(x, y)}{(c_n)_x} = 1 - \sum_{y \in V_m} \frac{c_n(x, y)}{(c_n)_x} = 0.$$

For $n > m$ we use the monotonicity of the quotients (see Corollary 4.35) and get

$$\sum_{y \notin V_m} \frac{c_n(x, y)}{(c_n)_x} = 1 - \sum_{y \in V_m} \frac{c_n(x, y)}{(c_n)_x} \leq 1 - \sum_{y \in V_m} \frac{c(x, y)}{c_x} < \varepsilon$$

Hence, we have proven that (T1) holds.

Now assume that (T1) holds. By Corollary 4.35 and (C), we have

$$c_x = \lim_{n \to \infty} \sum_{y \in V_n} c(x, y) \leq \lim_{n \to \infty} \sum_{y \in V_n} c_n(x, y) = \lim_{n \to \infty} (c_n)_x < \infty.$$

Let $\varepsilon > 0$. Then the convergence of the series $\sum_{y \in V} c(x, y)$ implies that there exists $m_1 \in \mathbb{N}$, such that

$$\sum_{y \notin V_{m_1}} c(x, y) < \frac{\varepsilon}{3}.$$

Furthermore, let $D_x := \lim_{n\to\infty}(c_n)_x$ and choose $m_2 \in \mathbb{N}$ such that

$$\sum_{y \notin V_{m_2}} \frac{c_n(x,y)}{(c_n)_x} < \frac{\varepsilon}{3D_x}$$

for all $n \in \mathbb{N}$. Now, let $m = \max(m_1, m_2)$ and let $n \in \mathbb{N}$ such that

$$\left| \sum_{y \in V_m} c_n(x,y) - c(x,y) \right| < \frac{\varepsilon}{3}.$$

This is possible due to the fact that $c_n(x,y) \to c(x,y)$ and that V_m is finite. It follows that

$$\left| \sum_{y \in V_n} c_n(x,y) - c(x,y) \right| \leq \left| \sum_{y \in V_m} c_n(x,y) - c(x,y) \right| + \left| \sum_{y \notin V_m} c_n(x,y) - c(x,y) \right|$$

$$< \frac{\varepsilon}{3} + \left| \sum_{y \notin V_{m_2}} c_n(x,y) \right| + \left| \sum_{y \notin V_{m_1}} c(x,y) \right|$$

$$< \frac{\varepsilon}{3} + (c_n)_x \cdot \frac{\varepsilon}{3D_x} + \frac{\varepsilon}{3} \leq \varepsilon.$$

The last inequality is due to Corollary 4.35 which implies that

$$D_x = \lim_{m \to \infty} (c_m)_x \geq (c_n)_x$$

for all $n \in \mathbb{N}$. Hence, we have shown that

$$\lim_{n\to\infty}(c_n)_x = \lim_{n\to\infty} \sum_{y \in V_n} c_n(x,y) = \lim_{n\to\infty} \sum_{y \in V_n} c(x,y) = \sum_{y \in V} c(x,y).$$

$\square$

Theorem 4.46. *Fix $x_0 \in V$. The sequence of random walks $(\mathbf{P}^n_{x_0})_{n \in \mathbb{N}}$ on (V_n, c_n) starting in x_0 is sequentially compact in the set of probability measures on $V^{\mathbb{N}_0}$ if and only if*

$$\sum_{y \in V} \lim_{n\to\infty} c_n(x,y) = \lim_{n\to\infty} \sum_{y \in V_n} c_n(x,y) < \infty \tag{4.20}$$

for all $x \in V$.

Proof. The claim follows by Prohorov's theorem, Lemma 4.44 and Lemma 4.45. $\square$

The following example shows that the condition (4.20) for tightness is not automatically satisfied and thus can not be omitted.

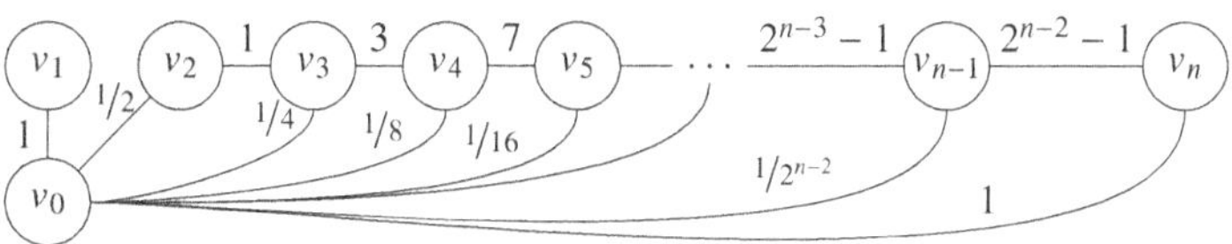

Figure 4.3: A sequence of graphs not satisfying condition (4.20).

Example 4.47. Consider the following sequence of graphs, denoted by $G_n = (V_n, c_n)$ where $V_n = \{v_0, \ldots, v_n\}$ and c_n is given by

$$c_n(v_0, v_k) = c_n(v_k, v_0) = \begin{cases} 1/2^{k-1} & , \ 1 \leq k < n \\ 1 & , \ k = n \end{cases},$$

$$c_n(v_k, v_{k+1}) = c(v_{k+1}, v_k) = 2^{k-1} - 1 \text{ for } 1 \leq k \leq n - 1$$

and $c_n(x, y) = 0$ otherwise, see Figure 4.3. Let R_n be the effective resistance of G_n. We verify that R_{n+1} agrees with R_n on V_n by applying the star-mesh transform (Proposition 2.6) to v_n in G_n. This yields a graph (V_{n-1}, c') with effective resistance $R_n \!\restriction_{V_{n-1}}$ and

$$c'(x, y) = c_n(x, y) + \frac{c_n(x, v_n) c_n(v_n, y)}{c_{v_n}}.$$

Hence, for all $x, y \in V_{n-1}$ such that $\{x, y\} \neq \{v_0, v_{n-1}\}$, we have $c'(x, y) = c_n(x, y) = c_{n-1}(x, y)$. Furthermore,

$$c'(v_0, v_{n-1}) = \frac{1}{2^{n-2}} + \frac{1 \cdot (2^{n-2} - 1)}{1 + 2^{n-2} - 1} = \frac{2^{n-2}}{2^{n-2}} = 1 = c_{n-1}(v_0, v_{n-1}).$$

It follows that $c' = c_{n-1}$ and thus $R_{n-1} = R_n \!\restriction_{V_{n-1}}$.

Now we will see that $(G_n)_{n \in \mathbb{N}}$ does not satisfy (4.20). We have

$$c(v_0, v_k) = \lim_{n \to \infty} c_n(v_0, v_k) = \frac{1}{2^{k-1}}$$

for all $k \in \mathbb{N}$ and thus

$$\sum_{y \in V} c(v_0, y) = \sum_{k=1}^{\infty} \frac{1}{2^{k-1}} = 2.$$

However,

$$\lim_{n \to \infty} (c_n)_{v_0} = \lim_{n \to \infty} \sum_{y \in V_n} c_n(v_0, y) = \lim_{n \to \infty} \left(1 + \sum_{k=1}^{n-1} \frac{1}{2^{k-1}} \right) = 3.$$

This concludes the example.

Remark 4.48. Note that Example 4.47 can be modified such that $c_n(v_0, v_n) = n$ and

$$c_n(v_k, v_{k+1}) = \frac{2^{k-1}(k+1)^2}{2^{k-1} + 1} - (k+1).$$

This will still yield a sequence of graphs with invariant effective resistances and $\sum_{y \in V} c(v_0, y) = 2$. However, in this case $\lim_{n \to \infty} (c_n)_{v_0} \geq n \to \infty$.

In the case of tightness, Prohorov's theorem only implies the existence of a weak limit point of $(\mathbf{P}_x^n)_{n \in \mathbb{N}}$. In our context, we can obtain the weak convergence of the whole sequence and identify the weak limit as random walk of the limit graph G_R.

Theorem 4.49. *If $c_x = \lim_{n \to \infty} (c_n)_x < \infty$ for all $x \in V$, then $(\mathbf{P}_x^n)_{n \in \mathbb{N}}$ converges weakly to the random walk $\mathbf{P}_x$ of the limit graph G_R for all $x \in V$.*

Proof. Let $x \in V$. By Theorem 4.46, $(\mathbf{P}_x^n)_{n \in \mathbb{N}}$ is tight. Hence, there exists a subsequence $(\mathbf{P}_x^{n_k})_{k \in \mathbb{N}}$ which weakly converges to a measure μ. For $x_0, \ldots, x_n \in V$, we have $A := \{x_0\} \times \ldots \times \{x_n\} \times V^{\mathbb{N}} \in C$. Hence, $\mathbf{1}_A$ is bounded and continuous and it follows that

$$\mu[X_0 = x_0, \ldots, X_n = x_n] = \int \mathbf{1}_A \, d\mu = \lim_{k \to \infty} \int \mathbf{1}_A \, d\mathbf{P}_x^{n_k}$$

$$= \lim_{n \to \infty} \mathbf{P}_x^{n_k}[X_0 = x_0, \ldots, X_n = x_n]$$

$$= \lim_{n \to \infty} \mathbf{1}_x(x_0) \cdot \prod_{l=0}^{n-1} \frac{c_{n_k}(x_l, x_{l+1})}{(c_{n_k})_{x_l}}$$

$$= \mathbf{1}_x(x_0) \cdot \prod_{l=0}^{n-1} \frac{c(x_l, x_{l+1})}{c_{x_l}} = \mathbf{P}_x[X_0 = x_0, \ldots, X_n = x_n].$$

Let $C \in C$. Then there exist $A_0, \ldots, A_n \subseteq V$ such that $C = A_0 \times \ldots \times A_n \times V^{\mathbb{N}}$. Note that $A_0 \times \ldots \times A_n$ is at most countably infinite. Since

$$C = \bigcup_{(x_0, \ldots, x_n) \in A_0 \times \ldots \times A_n} \{x_0\} \times \ldots \times \{x_n\} \times V^{\mathbb{N}}$$

and this is a disjoint union, the σ-additivity of μ and $\mathbf{P}_x$ implies that $\mu(C) = \mathbf{P}_x(C)$. Hence, μ and $\mathbf{P}_x$ agree on C. Since (Ω, ρ) is a countable product of separable metric spaces, its Borel-σ-algebra $\mathcal{B}$ satisfies

$$\mathcal{B} = \mathcal{B}(V) \otimes \mathcal{B}(V) \otimes \ldots = \sigma(C).$$

By (4.18), C is closed under finite intersections. It is a standard result that $\mu \restriction_C = \mathbf{P}_x \restriction_C$ implies $\mu = \mathbf{P}_x$.

Hence, we have shown that every weakly convergent subsequence of $(\mathbf{P}_x^n)_{n \in \mathbb{N}}$ converges to $\mathbf{P}_x$. Since every subsequence of a tight sequence is again tight, we obtain the following statement. Every subsequence of $(\mathbf{P}_x^n)_{n \in \mathbb{N}}$ contains a subsequence converging to $\mathbf{P}_x$. This is equivalent to $(\mathbf{P}_x^n)_{n \in \mathbb{N}}$ converging to $\mathbf{P}_x$. $\qquad\square$

By (2.17), we have for $x, y \in V_n$,

$$R(x, y) = \frac{1}{(c_n)_x} \int \sum_{k=0}^{\tau_y - 1} \mathbf{1}_x(X_k) \, d\mathbf{P}_x^n. \tag{4.21}$$

Since $(c_n)_x \to c_x$ and $\mathbf{P}_x^n$ converges weakly to $\mathbf{P}_x$, one may hope to get an analogous equation in terms of c_x and $\mathbf{P}_x$ which would then hold for all $x, y \in V$. The following example will show that this is in general false.

Figure 4.4: The graph G_n.

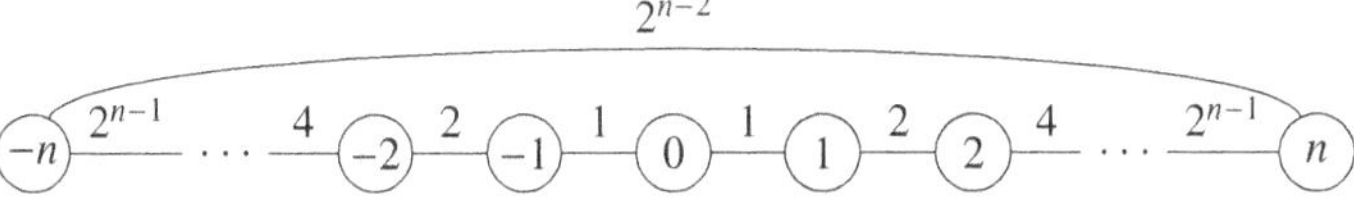

Figure 4.5: The graph H_n.

Example 4.50. We consider two sequences of graphs $G_n = (V_n, c_n)$ and $H_n = (V_n, \widehat{c_n})$ with $V_n = \{z \in \mathbb{Z} \mid |z| \le n\}$,

$$c_n(x, y) = \begin{cases} 0 & , |x - y| \ne 1 \\ 2^{\min(|x|,|y|)} & , |x - y| = 1 \end{cases},$$

see Figure 4.4 and $\widehat{c_n}(x, y) = c_n(x, y) + \mathbf{1}_{\{-n,n\}}(\{x, y\}) \cdot 2^{n-2}$, see Figure 4.5. Let $n \in \mathbb{N}$. Note that G_n and H_n are respectively obtained from G_{n+1} and H_{n+1} by applying the star-mesh transform (see Lemma 2.6) to the vertices $n + 1$ and $-(n + 1)$ successively. Hence, $R_{G_{n+1}}(x, y) = R_{G_n}(x, y)$ and $R_{H_{n+1}}(x, y) = R_{H_n}(x, y)$ for all $x, y \in V_n$. It follows that R_{G_n} and R_{H_n} are finite restrictions of resistance metrics R_1 and R_2 on V.

Since G_n is a tree, its effective resistance equals its geodesic metric. Let $f(x) := \sum_{k=0}^{|x|-1} 2^{-k}$. Then, for $x, y \in V_n$, we have

$$R_1(x, y) = R_{G_n}(x, y) = \begin{cases} f(x) + f(y) & , x \cdot y \le 0 \\ f(\max(|x|, |y|)) - f(\min(|x|, |y|)) & , x \cdot y > 0 \end{cases}.$$

Note that for $x > 0$, $R_1(x, n) = R_1(0, n) - R_1(0, x) = 2 - 2^{-(n-1)} - R_1(0, x)$.

Let $x, y \in V_n$ and w.l.o.g. assume that $x < y$. Using basic network reduction rules for parallel and series circuits, one obtains

$$R_2(x, y) = \left(\frac{1}{R_1(x, y)} + \frac{1}{R_1(x, -n) + 2^{-(n-2)} + R_1(y, n)} \right)^{-1}$$

$$= \left(\frac{1}{R_1(x, y)} + \frac{1}{4 - R_1(x, y)} \right)^{-1} = \frac{1}{4} R_1(x, y)(4 - R_1(x, y)).$$

Hence, R_1 and R_2 are different resistance metrics. However, both limit graphs are the same, namely $G = (\mathbb{Z}, c)$ where $c(x, y) = c_n(x, y)$ if $x, y \in V_n$. It follows that both sequences of random walks on G_n and on H_n have the same weak limit with respect to our chosen topology. Hence, it is impossible to write both R_1 and R_2 in terms of the random walk on G as in (4.21).

Note that R_1 and R_2 are the free and the wired effective resistance of G, respectively.

From now on assume that $\mathbf{P}_x^n$ converges weakly to $\mathbf{P}_x$. We write $\mathbf{P}_x^n \Rightarrow \mathbf{P}_x$. By definition, this gives us

$$\lim_{n\to\infty} \mathbf{E}_x^n[f] = \mathbf{E}_x[f]$$

for all continuous and bounded $f : \Omega \to \mathbb{R}$. Note that, by the Portmanteau theorem [11, Theorem 3.9.1], this can be extended to measurable and bounded $f : \Omega \to \mathbb{R}$ whose set of points of discontinuity is a null set of $\mathbf{P}_x$.

Recall Definition 2.17, namely

$$\Phi^{xy} = \sum_{k=0}^{\tau_y-1} \mathbf{1}_x(X_k).$$

Φ^{xy} is continuous on $\{\tau_y < \infty\}$ (but not on $\{\Phi^{xy} < \infty\}$) and not bounded. Hence, we can not directly apply weak convergence to obtain $\mathbf{E}_x^n[\Phi^{xy}] \to \mathbf{E}_x[\Phi^{xy}]$. To deal with the unboundedness, we first compute the distribution of Φ^{xy} with respect to $\mathbf{P}_x^n$.

As seen in the proof of Corollary 2.19, we have $\mathbf{P}_x^n[\Phi^{xy} = 0] = 0$ and, for all $k \in \mathbb{N}^+$,

$$\mathbf{P}_x^n[\Phi^{xy} = k] = (1 - p_n) \cdot p_n^{k-1}$$

where $p_n = \mathbf{P}_x^n[\tau_x^+ < \tau_y]$. Since we know the exact distribution of Φ^{xy}, we can compute its first and second moment with respect to $\mathbf{P}_x^n$. Because all graphs (V_n, c_n) are connected, we have $p_n < 1$ for all $n \in \mathbb{N}$. Hence, we see that

$$\mathbf{E}_x^n[\Phi^{xy}] = \sum_{k=1}^{\infty} k(1 - p_n)p_n^{k-1} = \frac{1}{1 - p_n} . \tag{4.22}$$

For the second moment, we have

$$\mathbf{E}_x^n\left[(\Phi^{xy})^2\right] = \sum_{k=1}^{\infty} k^2(1 - p_n)p_n^{k-1} = \frac{1 + p_n}{(1 - p_n)^2} . \tag{4.23}$$

Theorem 4.51. *Let R be a resistance metric on a countably infinite set V, $(V_n)_{n\in\mathbb{N}}$ a finite exhaustion of V, $G_n = (V_n, c_n)$ finite graphs with effective resistance $R\restriction_{V_n}$ and $G_R = (V, c)$ be the associated limit graph. If*

$$c_v = \lim_{n\to\infty} (c_n)_v < \infty$$

for all $v \in V$, and the random walk of G_R is recurrent, then for $x, y \in V$, we have

$$R(x, y) = \frac{1}{c_x}\mathbf{E}_x\left[\sum_{k=0}^{\tau_y-1} \mathbf{1}_x(X_k)\right].$$

Proof. By Theorem 4.49, we know that $\mathbf{P}_x^n \Rightarrow \mathbf{P}_x$. For simplicity, let $\Phi := \Phi^{xy}$. Throughout this proof assume that all n are large enough such that $x, y \in V_n$. By (4.21) and (4.22), we have

$$(c_n)_x \cdot R(x, y) = \mathbf{E}_x^n[\Phi] = \frac{1}{1 - p_n} .$$

Since $\lim_{n\to\infty}(c_n)_x < \infty$, it follows that p_n is uniformly bound away from 1 and thus

$$\mathbf{E}_x^n[\Phi^2] = \frac{1+p_n}{(1-p_n)^2} \le K$$

for some $K > 0$ and all sufficiently large n. Hence, for $T > 0$, we have

$$\int_{\{\Phi>T\}} \Phi \, d\mathbf{P}_x^n \le \frac{1}{T}\mathbf{E}_x^n[\Phi^2] \le \frac{K}{T}.$$

Let $\varepsilon > 0$. We define $\Phi^T(\omega) = \min(\Phi(\omega), T))$. Then, $\Phi^T \nearrow \Phi$ point-wise as $T \to \infty$. By the Theorem of monotone convergence, there exists T_1 such that

$$\mathbf{E}_x\left[\Phi - \Phi^T\right] \le \frac{\varepsilon}{3}$$

for all $T \ge T_1$. For $T \ge \max(T_1, 3K/\varepsilon)$, we then also have

$$\mathbf{E}_x^n\left[\Phi - \Phi^T\right] = \int_{\{\Phi>T\}} \Phi - T \, d\mathbf{P}_x^n \le \int_{\{\Phi>T\}} \Phi \, d\mathbf{P}_x^n \le \frac{K}{T} \le \frac{\varepsilon}{3}$$

for all n. Since $\mathbf{P}_x$ is recurrent, we have $\mathbf{P}_x[\tau_y < \infty] = 1$. Hence, Φ^T is continuous $\mathbf{P}_x$-almost everywhere and bounded. It follows that $\mathbf{E}_x^n[\Phi^T] \to \mathbf{E}_x[\Phi^T]$ and thus there exists $n_0 \in \mathbb{N}$ (dependent on T) such that

$$\left|\mathbf{E}_x\left[\Phi^T\right] - \mathbf{E}_x^n\left[\Phi^T\right]\right| \le \frac{\varepsilon}{3}$$

for all $n \ge n_0$. Hence,

$$\left|\mathbf{E}_x[\Phi] - \mathbf{E}_x^n[\Phi]\right| \le \left|\mathbf{E}_x[\Phi - \Phi^T]\right| + \left|\mathbf{E}_x[\Phi^T] - \mathbf{E}_x^n[\Phi^T]\right| + \left|\mathbf{E}_x^n[\Phi - \Phi^T]\right| \le \varepsilon.$$

It follows that $\mathbf{E}_x^n[\Phi] \to \mathbf{E}_x[\Phi]$ as $n \to \infty$. The claim follows from $\lim(c_n)_x = c_x$ and

$$R(x, y) = \frac{1}{(c_n)_x}\mathbf{E}_x^n\left[\Phi^{xy}\right]$$

by (2.17). $\qquad\square$

Remark 4.52. Combining Theorem 1.31 and Lemma 2.61 in [3], it follows that the (free) effective resistance of G_R equals

$$R_{G_R}(x, y) = \frac{1}{c_x \mathbf{P}_x[\tau_y \le \tau_x^+]} - \mathbf{E}_x\left[\sum_{k=0}^{\tau_y-1} \mathbf{1}_x(X_k)\right] \tag{4.24}$$

if G_R is recurrent. This differs from the above statement because it is not known that R is the effective resistance of G_R. However, it can be used to identify R as the (free) effective resistance of G_R.

Corollary 4.53. *Assume the same situation as in Theorem 4.51. If*

$$c_v = \lim_{n\to\infty}(c_n)_v < \infty$$

holds for all $v \in V$, and G_R is recurrent, then R is the effective resistance of G_R.

4.6 Free Effective Resistance

We will now see that the free effective resistance of an infinite graph $G = (V, c)$ can be realized as a resistance metric which is associated to the resistance form $(\mathcal{E}, \operatorname{dom} \mathcal{E})$. As in Section 3.2, let $G_n = G \restriction_{V_n}$ be the subgraph of G induced by the exhaustion V_n and $\mathcal{E}_n$ its energy form.

Lemma 4.54. *Let $f \in l(V)$, $f_n \in l(V_n)$ such that $f_n \to f$ point-wise. Then,*

$$\mathcal{E}(f) \le \liminf_{n \to \infty} \mathcal{E}_n(f_n). \tag{4.25}$$

Proof. For $n \in \mathbb{N}, x, y \in V$, let

$$F_n(x, y) := \mathbf{1}_{V_n}(x) \mathbf{1}_{V_n}(y) \cdot c(x, y)(f_n(x) - f_n(y))^2.$$

By Fatou's Lemma [34, Theorem 9.11], we have

$$\mathcal{E}(f) = \frac{1}{2} \sum_{x, y \in V} c(x, y)(f(x) - f(y))^2 = \frac{1}{2} \sum_{x, y \in V} \left(\lim_{n \to \infty} F_n(x, y) \right)$$

$$\le \liminf_{n \to \infty} \frac{1}{2} F_n(x, y) = \liminf_{n \to \infty} \mathcal{E}_n(f_n).$$

$\square$

Remark 4.55. While similar, this is not the same as the lower semi-continuity of $\mathcal{E}$ (Lemma 4.6). While our premise that $\mathcal{E}$ is an energy form of a graph is more restrictive, the result is also stronger as it implies lower semi-continuity (in the case of energy forms). Indeed, if $f_n \in l(V)$ converges point-wise to f, then $f_n \restriction_{V_n} \in l(V_n)$ does as well and we have

$$\mathcal{E}(f) \le \liminf_{n \to \infty} \mathcal{E}_n(f_n \restriction_{V_n}) \le \liminf_{n \to \infty} \mathcal{E}(f_n).$$

Theorem 4.56. *For any graph G, $(\mathcal{E}, \operatorname{dom} \mathcal{E})$ is a resistance form and we have*

$$R^F = R_{(\mathcal{E}, \operatorname{dom} \mathcal{E})}. \tag{4.26}$$

In particular,

$$R^F(x, y) = \max \left\{ \frac{(f(x) - f(y))^2}{\mathcal{E}(f)} \mid f \in \operatorname{dom} \mathcal{E}, \mathcal{E}(f) > 0 \right\} \tag{4.27}$$

$$= (\min \{ \mathcal{E}(f) \mid f \in \operatorname{dom} \mathcal{E}, f(x) = 1, f(y) = 0 \})^{-1}. \tag{4.28}$$

Proof. First we check that $(\mathcal{E}, \operatorname{dom} \mathcal{E})$ is a resistance form. Properties (RF-1) and (RF-2) are the statements of Lemma 1.13 and Proposition 1.16, respectively. For any finite $W \subseteq V$ and $f \in l(W)$, we have $\mathbf{1}_W \cdot f \in l_{\mathrm{fin}}(V) \subseteq \operatorname{dom} \mathcal{E}$ by Lemma 3.14. Thus, (RF-3) is satisfied. To see that (RF-4) holds, we proceed exactly as for resistance forms on finite sets (cf. Lemma 4.8) and get for all $x \in V$

$$0 < \frac{1}{c_x} = \frac{(\mathbf{1}_x(x) - \mathbf{1}_x(y))^2}{\mathcal{E}(\mathbf{1}_x)} \le R_{(\mathcal{E}, \operatorname{dom} \mathcal{E})}(x, y)$$

and, for any $f \in \mathrm{dom}\,\mathcal{E}$, $\mathcal{E}(f) > 0$, $x, y \in V$,

$$(f(x) - f(y))^2 \le \mathcal{E}(f) \cdot d_G(x, y).$$

Since we have

$$|m(f)(x) - m(f)(y)| \le |f(x) - f(y)|$$

for all $x, y \in V$, $\mathcal{E}(m(f)) < \mathcal{E}(f)$ follows. Hence, (RF-5) holds and thus $(\mathcal{E}, \mathrm{dom}\,\mathcal{E})$ is a resistance form.

We now show that $R^F = R_{(\mathcal{E}, \mathrm{dom}\,\mathcal{E})}$. Let $x, y \in V$, $x \ne y$ and, for $n \in \mathbb{N}$, let

$$\psi_n := \mathrm{argmin}\,\{\mathcal{E}_n(f) \mid f \in l(V_n), f(x) = 1, f(y) = 0\}\,.$$

Then,

$$R_{G_n}(x, y)^{-1} = \mathcal{E}_n(\psi_n) \le \mathcal{E}_n\left(\psi^{xy}_{\mathrm{dom}\,\mathcal{E}} \upharpoonright V_n\right) \le \mathcal{E}\left(\psi^{xy}_{\mathrm{dom}\,\mathcal{E}}\right) = R_{(\mathcal{E}, \mathrm{dom}\,\mathcal{E})}(x, y)^{-1}$$

and it follows that $R^F(x, y) \ge R_{(\mathcal{E}, \mathrm{dom}\,\mathcal{E})}(x, y)$.

Suppose we know that there is a subsequence $(\psi_{n_k})_{k \in \mathbb{N}}$ such that, for all $v \in V$, $\widetilde{\psi}(v) := \lim_{k \to \infty} \psi_{n_k}(v)$ exists. Then, $\widetilde{\psi}(x) = 1$, $\widetilde{\psi}(y) = 0$ and, by (4.25), we have

$$R_{(\mathcal{E}, \mathrm{dom}\,\mathcal{E})}(x, y)^{-1} = \mathcal{E}\left(\psi^{xy}_{\mathrm{dom}\,\mathcal{E}}\right) \le \mathcal{E}(\widetilde{\psi}) \le \liminf_{k \to \infty} \mathcal{E}_{n_k}(\psi_{n_k}) = R^F(x, y)^{-1}$$

and the claim follows.

We now show the existence of such a subsequence. Let $V = \{v_1, v_2, \ldots\}$ such that $v_k \in V_k$ for all $k \in \mathbb{N}$. For every $n \in \mathbb{N}$, we inductively define a sequence $(i(n, k))_{k \in \mathbb{N}}$ of strictly increasing natural numbers as follows. Since ψ_n is a minimizer of $\mathcal{E}_n$ with boundary values $\psi_n(x) = 1$, $\psi_n(y) = 0$, it follows by (RF-5) that $0 \le \psi_n \le 1$. Hence, there exists a subsequence $(i(1, k))_{k \in \mathbb{N}}$ of $(1, 2, \ldots)$ such that $\lim_{k \to \infty} \psi_{i(1,k)}(v_1)$ exists. For $n > 1$, we use the same argument to define $i(n, k)$ as a subsequence of $i(n - 1, k)$ such that $\lim_{k \to \infty} \psi_{i(n,k)}(v_n)$ exists. Then, $\lim_{k \to \infty} \psi_{i(n,k)}(v)$ exists for all $v \in \{v_1, \ldots, v_n\}$ by induction.

Let $\widetilde{\psi}(v_n) := \lim_{k \to \infty} \psi_{i(n,k)}(v_n)$. For $k \ge n$, $i(k, k)$ is a subsequence of $i(n, k)$. Hence,

$$\lim_{k \to \infty} \psi_{i(k,k)}(v_n) = \lim_{k \to \infty} \psi_{i(n,k)}(v_n) = \widetilde{\psi}(v_n)$$

follows. Therefore, we have found the desired subsequence $n_k := i(k, k)$, $k \in \mathbb{N}$. $\qquad\square$

Remark 4.57. In [19, Remark 2.24] it is stated that R^F is the resistance metric associated to $(\mathcal{E}, \mathrm{dom}\,\mathcal{E})$ since

$$R^F(x, y) = (\min\,\{\mathcal{E}(u) \mid u(x) = 1,\ u(y) = 0,\ f \in \mathrm{dom}\,\mathcal{E}\})^{-1}\,.$$

However, the proof of this equation (see [19, Theorem 2.14]) seems to be incomplete. Indeed, Jorgensen and Pearse make the argument that

$$\mathcal{E}(v_{xy}) = \lim_{k \to \infty} \sum_{s,t \in V_k} c(s, t) \cdot (v_{xy}(s) - v_{xy}(t))^2 = \lim_{k \to \infty} R_{G_k}(x, y)$$

where v_{xy} corresponds to a re-scaled version of our potential ϕ^{xy}. However, they seem to assume that the unit potentials on G_k are simply restrictions of ϕ^{xy} to V_k which is not the case.

4.7   Wired Effective Resistance

The wired effective resistance of an infinite graph can also be realized as a resistance metric. Its associated resistance form is $(\mathcal{E}, \mathrm{Fin})$. Let V_n^W be a wired exhaustion and G_n^W the induced finite graphs. As in Section 3.3, we denote its energy form and Laplacian by $\mathcal{E}_n^W$ and Δ_n^W, respectively. Recall that $L_n^W : l(V_n^W) \to l_{\mathrm{fin}}(V; V_n)$ is a bijection satisfying $\mathcal{E}(L_n^W(f)) = \mathcal{E}_n^W(f)$. In particular, we have

$$\min_{\substack{f \in l(V_n^W) \\ f(x)=1, f(y)=0}} \mathcal{E}_n^W(f) = \min_{\substack{f \in l_{\mathrm{fin}}(V;V_n) \\ f(x)=1, f(y)=0}} \mathcal{E}(f).$$

Theorem 4.58. *For any graph G, $(\mathcal{E}, \mathrm{Fin})$ is a resistance form and we have*

$$R^W = R_{(\mathcal{E},\mathrm{Fin})}. \tag{4.29}$$

In particular,

$$R^W(x, y) = \max \left\{ \frac{(f(x) - f(y))^2}{\mathcal{E}(f)} \mid f \in \mathrm{Fin}, \mathcal{E}(f) > 0 \right\} \tag{4.30}$$

$$= (\min \{\mathcal{E}(f) \mid f \in \mathrm{Fin}, f(x) = 1, f(y) = 0\})^{-1}. \tag{4.31}$$

Proof. Since Fin is a closed subspace of $\mathrm{dom}\,\mathcal{E}$, Theorem 4.56 implies that (RF-1) and (RF-2) are satisfied. Property (RF-3) holds since $l_{\mathrm{fin}}(V) \subseteq \mathrm{Fin}$ by definition of Fin. Since G is locally finite, we have $\mathbf{1}_x \in \mathrm{Fin}$ for all $x \in V$ and thus (RF-4) holds by the same arguments as in the proof of Theorem 4.56.

In order to show (RF-5), let $f \in \mathrm{Fin}$. Then, there exists $k \in \mathbb{R}$ and $f_n \in l_{\mathrm{fin}}(V)$ such that $f_n + k \to f$ in H_x, i.e. $f_n + k \to f$ point-wise and $\mathcal{E}(f_n + k - f) = \mathcal{E}(f_n - f) \to 0$. It follows that $m(f_n + k) \in l_{\mathrm{fin}}(V)$ and $m(f_n + k) \to m(f)$ point-wise. We have to show that $\mathcal{E}(m(f_n + k) - m(f)) \to 0$ holds as well. Note that while

$$\forall\, a, b \in \mathbb{R} : |m(a) - m(b)| \le |a - b|$$

holds,

$$\forall\, a, b, c, d \in \mathbb{R} : |m(a) - m(b) - (m(c) - m(d))| \le |a - b - (c - d)|$$

is **not** true (take e.g. $a = -1/2$, $b = -1$, $c = 1$, $d = 1/2$). Hence, while $\mathcal{E}(m(f)) \le \mathcal{E}(f)$ holds true, $\mathcal{E}(m(f_n + k) - m(f)) \le \mathcal{E}(f_n + k - f)$ is generally false.

Instead, we will utilize Vitali's convergence theorem (Theorem 1.28) in the measure space $(V \times V, 2^{V \times V}, \mu)$ where $\mu(\{(x, y)\}) = c(x, y)/2$. Then, μ is σ-finite and for any function $F : V \times V \to \mathbb{R}$, we have

$$\int F \, d\mu = \frac{1}{2} \sum_{x,y \in V} c(x, y) F(x, y).$$

Now let $F(x, y) := f(x) - f(y)$ and $F_n(x, y) := f_n(x) - f_n(y)$. It follows that

$$\int |F_n - F|^2 \, d\mu = \mathcal{E}(f_n - f) \to 0$$

and Vitali's convergence theorem implies that $(|F_n|^2)_{n \in \mathbb{N}}$ is uniformly integrable. Let $M(x, y) := (m(f)(x) - m(f)(y))$ and $M_n(x, y) := (m(f_n + k)(x) - m(f_n + k)(y))$. Then,

$$|M_n(x, y)| = |m(f_n + k)(x) - m(f_n + k)(y)| \leq |f_n(x) - f_n(y)| = |F_n(x, y)|$$

for all $x, y \in V$. Since $(|F_n|^2)_{n \in \mathbb{N}}$ is uniformly integrable, for any $\varepsilon > 0$ there exists $w_\varepsilon \in \mathcal{L}^1_+(\mu)$ such that

$$\sup_{n \in \mathbb{N}} \int_{\{|M_n|^2 > w_\varepsilon\}} |M_n|^2 \, d\mu \leq \sup_{n \in \mathbb{N}} \int_{\{|F_n|^2 > w_\varepsilon\}} |F_n|^2 \, d\mu < \varepsilon.$$

Hence, $(|M_n|^2)_{n \in \mathbb{N}}$ is uniformly integrable as well and Vitali's convergence theorem implies that $M_n \to M$ in $\mathcal{L}^2(\mu)$. Therefore,

$$\mathcal{E}(m(f_n + k) - m(f)) = \int |M_n - M|^2 \, d\mu \to 0$$

and it follows that $m(f) \in \mathrm{Fin}$. Thus, we have shown (RF-5).

Now we show that $R^W = R_{(\mathcal{E}, \mathrm{Fin})}$. Let $x, y \in V$, $x \neq y$ and

$$w_n^{xy} = \mathrm{argmin}\left\{\mathcal{E}_n^W(f) \mid f \in l(V_n^W), f(x) = 1, f(y) = 0\right\}.$$

Then, we have $R^W(x, y) = R_{\mathrm{Fin}}(x, y)$ if and only if

$$\mathcal{E}(\psi_{\mathrm{Fin}}^{xy}) = \lim_{n \to \infty} \mathcal{E}_n^W(w_n^{xy}).$$

Since $l_{\mathrm{fin}}(V; V_n) \subseteq l_{\mathrm{fin}}(V; V_{n+1}) \subseteq \mathrm{Fin}$ holds, we have

$$\min_{\substack{f \in l_{\mathrm{fin}}(V; V_n) \\ f(x)=1, f(y)=0}} \mathcal{E}(f) \geq \min_{\substack{f \in l_{\mathrm{fin}}(V; V_{n+1}) \\ f(x)=1, f(y)=0}} \mathcal{E}(f) \geq \min_{\substack{f \in \mathrm{Fin} \\ f(x)=1, f(y)=0}} \mathcal{E}(f).$$

Together with
$$\mathcal{E}_n^W(w_n^{xy}) = \mathcal{E}(L_n^W(w_n^{xy})) = \min_{\substack{f \in l_{\mathrm{fin}}(V; V_n) \\ f(x)=1, f(y)=0}} \mathcal{E}(f),$$

it follows that
$$\mathcal{E}_n^W(w_n^{xy}) \geq \mathcal{E}_{n+1}^W(w_{n+1}^{xy}) \geq \mathcal{E}(\psi_{\mathrm{Fin}}^{xy})$$

for all $n \in \mathbb{N}$. Hence, $\lim_{n \to \infty} \mathcal{E}_n^W(w_n^{xy})$ exists and is bounded from below by $\mathcal{E}(\psi_{\mathrm{Fin}}^{xy})$.

Since $\psi_{\mathrm{Fin}}^{xy} \in \mathrm{Fin}$, there exist finite sets $S_n \subseteq V$, $\psi_n \in l_{\mathrm{fin}}(V; S_n)$, such that $\psi_n \to \psi_{\mathrm{Fin}}^{xy}$ point-wise and w.r.t. $\mathcal{E}$. Since $\psi_n(x) \to 1$ and $\psi_n(y) \to 0$, the function $\widetilde{\psi_n} : V \to \mathbb{R}$,

$$\widetilde{\psi_n}(z) := \frac{\psi_n(z) - \psi_n(y)}{\psi_n(x) - \psi_n(y)}, \ z \in V$$

is well-defined for n sufficiently large and also satisfies $\widetilde{\psi_n} \in l_{\mathrm{fin}}(V; S_n)$.

Let $k(0) = \inf\left\{j \geq 0 \mid S_0 \subseteq V_j\right\}$ and for $n \geq 0$, let

$$k(n + 1) := \inf\left\{j > k(n) \mid S_{n+1} \subseteq V_j\right\}.$$

Then, $k(n) \to \infty$ as $n \to \infty$, $\widetilde{\psi_n} \in l_{\mathrm{fin}}(V; V_{k(n)})$, $\widetilde{\psi_n}(x) = 1$, $\widetilde{\psi_n}(y) = 0$ and we have

$$\mathcal{E}(\psi^{xy}_{\mathrm{Fin}}) = \lim_{n \to \infty} \frac{\mathcal{E}(\psi_n)}{(\psi_n(x) - \psi_n(y))^2}$$
$$= \lim_{n \to \infty} \mathcal{E}\left(\widetilde{\psi_n}\right) \geq \lim_{n \to \infty} \mathcal{E}(L^W_{k(n)}(w^{xy}_{k(n)})) = \lim_{n \to \infty} \mathcal{E}^W_n(w^{xy}_n).$$

$$\square$$

Remark 4.59. Similar to the case of the free effective resistance (Theorem 4.56), it is mentioned in [19, Remark 2.24] that R^W is the resistance metric associated to $(\mathcal{E}, \mathrm{Fin})$. However, the proof contained in [19] only pertains to the value of R^W, not the fact that $(\mathcal{E}, \mathrm{Fin})$ indeed is a resistance form.

In [25, Proposition 2.5], the **minimal resistance** is defined by directly adding a "point at infinity" to V and showing that this yields a resistance form on this extended space which is a natural extension of $\mathcal{E}$. It is mentioned that the associated resistance metric is sometimes called wired resistance in other literature.

4.8 Resistance Metrics of Graphs

In the preceding sections, we have seen that Kigami's theory of resistance forms is very general as it allows the definition of resistance metrics on arbitrary sets. We have also seen that it encompasses the notions of free and wired effective resistance. In this section, we will restrict the theory to the case when the resistance form "comes from a graph". This gives rise to a whole spectrum of resistance metrics of which the wired and free effective resistance are the extreme points.

Definition 4.60. Let $G = (V, c)$ be a graph with energy form $\mathcal{E}$. We say a mapping $R : V \times V \to \mathbb{R}$ is a **resistance metric of** G if there exists a linear space $\mathrm{Fin} \subseteq D \subseteq \mathrm{dom}\,\mathcal{E}$ such that $R = R_{(\mathcal{E},D)}$. In this case, we call $(\mathcal{E}, D)$ a **resistance form of** G.

When comparing different resistance metrics of the same graph, we will simply write R_D instead of $R_{(\mathcal{E},D)}$.

By definition and theorems 4.56 and 4.58, it immediately follows that

$$R^W(x, y) \leq R_D(x, y) \leq R^F(x, y) \tag{4.32}$$

for all $x, y \in V$. We also observe that the definition of a limit graph from the previous section is indeed a natural one as it reproduces G when applied to a resistance metric of G.

Proposition 4.61. *Let $G = (V, c)$ be a countably infinite graph with energy form $\mathcal{E}$ and let R be a resistance metric of G. Then, G is the limit graph of R.*

Proof. Let $R = R_{(\mathcal{E},D)}$ with $\mathrm{Fin} \subseteq D \subseteq \mathrm{dom}\,\mathcal{E}$. By (1.35), for $x, y \in V$, $x \neq y$, we have

$$c(x, y) = -\mathcal{E}(\mathbf{1}_x, \mathbf{1}_y).$$

Let $(V_n)_{n\in\mathbb{N}}$ be a finite exhaustion of V. Furthermore, let $G_n = (V_n, c_n)$ be the unique graph such that $R_{G_n} = R\restriction_{V_n}$. By Lemma 4.9 and the uniqueness of G_n, it follows that $\mathcal{E}_{G_n}$ is the trace of $(\mathcal{E}, D)$ on V_n, i.e. $\mathcal{E}_{G_n} = \mathcal{E}^{V_n}$. Fix $x, y \in V$, $x \neq y$. By Lemma 4.23, we have

$$c(x, y) = -\mathcal{E}(\mathbf{1}_x, \mathbf{1}_y) = \lim_{n\to\infty}\left(-\mathcal{E}^{V_n}(\mathbf{1}_x, \mathbf{1}_y)\right)$$
$$= \lim_{n\to\infty}\left(-\mathcal{E}_{G_n}(\mathbf{1}_x, \mathbf{1}_y)\right) = \lim_{n\to\infty} c_n(x, y) = c_R(x, y)$$

Hence, $c_R = c$ and therefore $G = G_R$. $\qquad\square$

Below, we reproduce and extend a well-known characterization of when a graph satisfies $R^W = R^F$, see e.g. [9, 30]. Such a graph is often described as having **unique currents**.

Proposition 4.62. *For a graph G, the following are equivalent:*

(i) There exists only one resistance metric of G

(ii) $R^F = R^W$

(iii) Fin = dom $\mathcal{E}$

(iv) Harm $\simeq \mathbb{R}$

In particular, recurrence of G implies the above statements.

Proof. By theorems 4.56 and 4.58, $R^F = R_{\mathrm{dom}\,\mathcal{E}}$ and $R^W = R_{\mathrm{Fin}}$ are resistance metric of G. Hence, $(i) \Rightarrow (ii)$ holds. If $R_{\mathrm{dom}\,\mathcal{E}} = R^F = R^W = R_{\mathrm{Fin}}$, then Corollary 4.24 implies Fin = dom $\mathcal{E}$. $(iii) \Rightarrow (i)$ holds by definition. The Royden decomposition (Theorem 3.18) implies $(iii) \Leftrightarrow (iv)$.

Lastly, Corollary 3.20 states that recurrence of G implies $l_0(V) = \mathrm{dom}\,\mathcal{E}$. In particular, Fin = dom $\mathcal{E}$ follows. $\qquad\square$

Remark 4.63. The converse statement that $R^F = R^W$ implies recurrence of G is **not** true, see Example 5.11.

In the following, let $G = (V, c)$ be a graph and R_D a resistance metric of G. For $x, y \in V$, $x \neq y$, define

$$D^{xy} := \{f \in D \mid f(x) = f(y) = 0\}$$

and

$$H_1^{xy} := \{h \in \mathrm{Harm} \mid h(x) = 1, h(y) = 0\}.$$

Since $\mathbf{1}_x \in \mathrm{Fin} \subsetneq D$ for all $x \in V$, we have

$$D_1^{xy} := \mathbf{1}_x + D^{xy} = \{f \in D \mid f(x) = 1, f(y) = 0\}.$$

By Corollaries 4.12 and 4.13, there exist ψ_D^{xy} and ϕ_D^{xy} such that $\psi_D^{xy}(x) = 1$, $\psi_D^{xy}(y) = 0$, $\mathcal{E}(\psi_D^{xy}) = R_D(x, y)^{-1}$ and $\phi_D^{xy}(x) = R_D(x, y)$, $\phi_D^{xy}(y) = 0$, $\mathcal{E}(\phi_D^{xy}) = R_D(x, y)$. Furthermore, we have $\phi_D^{xy} = \psi_D^{xy}/\mathcal{E}(\psi_D^{xy})$. We will now investigate the connection between those functions for different resistance metrics.

Proposition 4.64. *All potentials of $\mathcal{E}$ are dipoles, i.e.*

$$\Delta \phi_D^{xy} = \frac{1}{c_x}\mathbf{1}_x - \frac{1}{c_y}\mathbf{1}_y \, .$$

Furthermore, we have

$$\Delta \psi_D^{xy} = \frac{1}{R_D(x,y)} \cdot \left(\frac{1}{c_x}\mathbf{1}_x - \frac{1}{c_y}\mathbf{1}_y \right).$$

Proof. By (1.33), we have $\mathcal{E}(f, \mathbf{1}_v) = c_v(\Delta f)(v)$ for all $v \in V$, $f \in \mathrm{dom}\,\mathcal{E}$. Hence, Proposition 4.16 implies

$$(\Delta \phi_D^{xy})(z) = \frac{1}{c_z} \cdot \mathcal{E}(\phi_D^{xy}, \mathbf{1}_z) = \frac{1}{c_x}\mathbf{1}_x(z) - \frac{1}{c_y}\mathbf{1}_y(z)$$

for all $z \in V$. The second claim follows by

$$\psi_D^{xy} = \mathcal{E}(\psi_D^{xy}) \cdot \phi_D^{xy} = \frac{\phi_D^{xy}}{R_D(x,y)} \, .$$

$\square$

The above statement again shows that the theory of resistance metrics is a natural extensions of effective resistances on finite graphs.

By the Markov property of $\mathcal{E}$, we know that ψ_D^{xy} has values in $[0, 1]$. Hence, ϕ_D^{xy} has values in $[0, R_D(x,y)]$ and both functions attain their maximum and minimum at x and y, respectively.

Lemma 4.65. *If $H_1^{xy} \cap D = \emptyset$, then $\psi_D^{xy} = \psi_{\mathrm{Fin}}^{xy}$.*

Proof. Note that $H_1^{xy} \cap D = \emptyset$ implies $h(x) = h(y)$ for all $h \in \mathrm{Harm} \cap D$.

By the Royden decomposition of H_x (Cor. 3.21), we have $\psi_D^{xy} = f + h$ for some $f \in \mathrm{Fin}$, $h \in \mathrm{Harm}$. Since $f \in \mathrm{Fin} \subsetneq D$, it follows that $h \in \mathrm{Harm} \cap D$ and thus $h(x) = h(y) =: k$. Hence, $f(x) = \psi_D^{xy}(x) - h(x) = 1 - k$ and $f(y) = -k$. Since $f + k \in \mathrm{Fin}_1^{xy} \subsetneq D_1^{xy}$, we have

$$\mathcal{E}(\psi_D^{xy}) \leq \mathcal{E}(\psi_{\mathrm{Fin}}^{xy}) \leq \mathcal{E}(f + k) = \mathcal{E}(f) \leq \mathcal{E}(f) + \mathcal{E}(h) = \mathcal{E}(\psi_D^{xy}).$$

The uniqueness of ψ_D^{xy} and ψ_{Fin}^{xy} implies $\psi_D^{xy} = f + k = \psi_{\mathrm{Fin}}^{xy}$. $\square$

Lemma 4.66. *If $H_1^{xy} \cap D \neq \emptyset$, there exist $h_D \in H_1^{xy} \cap D$, $f_D \in \mathrm{Fin}_1^{xy}$ and $\lambda \in \mathbb{R}$ such that*

$$\psi_D^{xy} = \lambda \cdot f_D + (1 - \lambda) \cdot h_D.$$

Proof. For convenience, write $\psi = \psi_D^{xy}$. By the Royden decomposition, we always have $\psi = f + h$ for some $h \in \mathrm{Harm}$, $f \in \mathrm{Fin}$.

First, suppose that $h(x) = h(y) =: k$. As in the proof of Lemma 4.65, $\psi = f + k =: f_D$ follows and the claim is true for $\lambda = 1$ and any $h_D \in H_1^{xy} \cap D$.

Secondly, suppose that $f(x) = f(y) =: k$. Analogously to the first case, the claim follows for $\lambda = 0$, $h_D := h + k \in H_1^{xy} \cap D$ and any f_D.

Lastly, assume that $h(x) \neq h(y)$ and $f(x) \neq f(y)$. Define

$$f_D := \frac{f - f(y)}{f(x) - f(y)} \in \mathrm{Fin}_1^{xy}$$

and

$$h_D := \frac{h - h(y)}{h(x) - h(y)} \in H_1^{xy} \cap D.$$

Then, for $\lambda := f(x) - f(y)$ we have

$$1 - \lambda = \psi(x) - \psi(y) - (f(x) - f(y)) = h(x) - h(y)$$

and

$$0 = \psi(y) = f(y) + h(y).$$

Hence,

$$\lambda \cdot f_D + (1 - \lambda) \cdot h_D = f - f(y) + h - h(y) = f + h = \psi.$$

$\square$

Lemma 4.67. *For any* $f \in \mathrm{Fin}$, $h \in \mathrm{Harm}$, *we have*

$$\inf_{\lambda \in \mathbb{R}} \mathcal{E}(\lambda \cdot f + (1 - \lambda) \cdot h) = \left(\frac{1}{\mathcal{E}(f)} + \frac{1}{\mathcal{E}(h)} \right)^{-1}$$

which is attained at $\lambda_0 = \mathcal{E}(h)/(\mathcal{E}(f) + \mathcal{E}(h))$.

Proof. We have

$$E(\lambda) := \mathcal{E}(\lambda f + (1 - \lambda)h) = \lambda^2 \mathcal{E}(f) + (1 - \lambda)^2 \mathcal{E}(h).$$

Hence, $E(\lambda) \to \infty$ as $\lambda \to \pm\infty$. The derivations of $E(\lambda)$ w.r.t. λ are

$$\frac{\mathrm{d}}{\mathrm{d}\lambda} E(\lambda) = 2\lambda\mathcal{E}(f) - 2(1 - \lambda)\mathcal{E}(h) \text{ and}$$

$$\frac{\mathrm{d}^2}{\mathrm{d}^2\lambda} E(\lambda) = 2\mathcal{E}(f) + 2\mathcal{E}(h) > 0.$$

Hence, $E(\lambda)$ has a local minimum at $\lambda_0 = \mathcal{E}(h)/(\mathcal{E}(f) + \mathcal{E}(h))$ and its value is

$$E(\lambda_0) = \frac{\mathcal{E}(h)^2\mathcal{E}(f)}{(\mathcal{E}(f) + \mathcal{E}(h))^2} + \frac{\mathcal{E}(f)^2\mathcal{E}(h)}{(\mathcal{E}(f) + \mathcal{E}(h))^2} = \frac{\mathcal{E}(f)\mathcal{E}(h)}{\mathcal{E}(f) + \mathcal{E}(h)} = \left(\frac{1}{\mathcal{E}(f)} + \frac{1}{\mathcal{E}(h)} \right)^{-1}.$$

$\square$

Theorem 4.68. *Assume that* $H_1^{xy} \cap D \neq \emptyset$. *For* $x, y \in V$, $x \neq y$, *let* h_D^{xy} *be a minimizer of* $\mathcal{E}$ *in* $H_1^{xy} \cap D$. *Then,*

$$\psi_D^{xy} = \frac{\mathcal{E}(h_D^{xy})}{\mathcal{E}(\psi_{\mathrm{Fin}}^{xy} + h_D^{xy})}\psi_{\mathrm{Fin}}^{xy} + \frac{\mathcal{E}(\psi_{\mathrm{Fin}}^{xy})}{\mathcal{E}(\psi_{\mathrm{Fin}}^{xy} + h_D^{xy})}h_D^{xy} \tag{4.33}$$

and

$$\phi_D^{xy} = \phi_{\mathrm{Fin}}^{xy} + \frac{h_D^{xy}}{\mathcal{E}(h_D^{xy})} . \tag{4.34}$$

Proof. By Lemma 4.66 it follows that

$$\psi_D^{xy} = \mathrm{argmin}\left\{\mathcal{E}(u) \mid u = \lambda f + (1-\lambda)h, f \in \mathrm{Fin}_1^{xy}, h \in H_1^{xy} \cap D\right\}.$$

Applying Lemma 4.67, yields

$$\mathcal{E}(\psi_D^{xy}) = \min\left\{\left(\frac{1}{\mathcal{E}(f)} + \frac{1}{\mathcal{E}(h)}\right)^{-1} \mid f \in \mathrm{Fin}_1^{xy}, h \in H_1^{xy} \cap D\right\} =: M.$$

The minimum on the right-hand side is attained when minimizing both energies separately. Since

$$\psi_{\mathrm{Fin}}^{xy} = \mathrm{argmin}_{f \in \mathrm{Fin}_1^{xy}} \mathcal{E}(f)$$

and

$$\mathcal{E}(h_D^{xy}) = \min_{h \in H_1^{xy} \cap D} \mathcal{E}(h),$$

we have

$$\mathcal{E}(\psi_D^{xy}) = M = \left(\frac{1}{\mathcal{E}(\psi_{\mathrm{Fin}}^{xy})} + \frac{1}{\mathcal{E}(h_D^{xy})}\right)^{-1} = \mathcal{E}(\psi').$$

where

$$\psi' = \frac{\mathcal{E}(h_D^{xy})}{\mathcal{E}(\psi_{\mathrm{Fin}}^{xy} + h_D^{xy})}\psi_{\mathrm{Fin}}^{xy} + \frac{\mathcal{E}(\psi_{\mathrm{Fin}}^{xy})}{\mathcal{E}(\psi_{\mathrm{Fin}}^{xy} + h_D^{xy})}h_D^{xy}.$$

Since $\psi' \in D_1^{xy}$, uniqueness of ψ_D^{xy} implies $\psi_D^{xy} = \psi'$.

By definition of ϕ_D^{xy}, we have

$$\phi_D^{xy} = \frac{\psi_D^{xy}}{\mathcal{E}(\psi_D^{xy})} = \frac{\mathcal{E}(\psi_{\mathrm{Fin}}^{xy} + h_D^{xy})}{\mathcal{E}(\psi_{\mathrm{Fin}}^{xy})\mathcal{E}(h_D^{xy})} \cdot \left(\frac{\mathcal{E}(h_D^{xy})}{\mathcal{E}(\psi_{\mathrm{Fin}}^{xy} + h_D^{xy})}\psi_{\mathrm{Fin}}^{xy} + \frac{\mathcal{E}(\psi_{\mathrm{Fin}}^{xy})}{\mathcal{E}(\psi_{\mathrm{Fin}}^{xy} + h_D^{xy})}h_D^{xy}\right)$$

$$= \frac{\psi_{\mathrm{Fin}}^{xy}}{\mathcal{E}(\psi_{\mathrm{Fin}}^{xy})} + \frac{h_D^{xy}}{\mathcal{E}(h_D^{xy})} = \phi_D^{xy} + \frac{h_D^{xy}}{\mathcal{E}(h_D^{xy})}$$

$\square$

Corollary 4.69. *For $x, y \in V$, we have $R_D(x,y) \neq R^W(x,y)$ if and only if $\mathrm{Harm} \cap D$ separates the points x and y. In that case,*

$$R_D(x,y) = R^W(x,y) + \mathcal{E}(h_D^{xy})^{-1} \tag{4.35}$$

where h_D^{xy} is a minimizer of $\mathcal{E}$ in $H_1^{xy} \cap D$.

Proof. $\mathrm{Harm} \cap D$ separates $x, y \in V$ if and only if there exists $h \in \mathrm{Harm} \cap D$ such that $h(x) \neq h(y)$. Then,

$$\widetilde{h} := \frac{h - h(y)}{h(x) - h(y)}$$

satisfies $h(x) = 1$ and $h(y) = 0$. Hence, $H_1^{xy} \cap D \neq \emptyset$ and, by Theorem 4.68, we have

$$R_D(x,y) = \phi_D^{xy}(x) = \phi_{\mathrm{Fin}}^{xy}(x) + \frac{h_D^{xy}(x)}{\mathcal{E}(h_D^{xy})} = R^W(x,y) + \frac{1}{\mathcal{E}(h_D^{xy})}.$$

$\square$

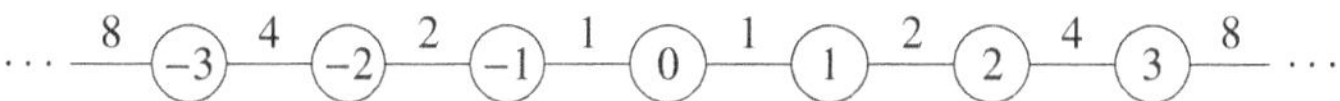

Figure 4.6: The network $(\mathbb{Z}, c)$.

Corollary 4.70. *For $x, y \in V$, $x \neq y$, we have either $H_1^{xy} = \emptyset$ or*

$$(R^F(x, y) - R^W(x, y))^{-1} = \min_{h \in H_1^{xy}} \mathcal{E}(h). \tag{4.36}$$

Example 4.71. Consider the network $(\mathbb{Z}, c)$ from our examples in Chapter 3, see Figure 4.6. We will now compute $\min_{h \in H_1^{xy}} \mathcal{E}(h)$ for $x = 1$ and $y = -1$. One can easily verify, that there exists exactly one harmonic function $h_0 \in l(\mathbb{Z})$ such that $h_0(1) = 1$ and $h_0(-1) = 0$. In fact, it is given by

$$h_0(z) = \frac{1}{2}(f(z) + 1) = \text{sign}(z) \cdot (1 - 2^{-|z|}) + \frac{1}{2}$$

where f is taken from Example 3.1. In that example we also computed $\mathcal{E}(f) = 4$ which implies

$$\mathcal{E}(h_0) = \mathcal{E}\left(\frac{1}{2}f + \frac{1}{2}\right) = \frac{1}{4}\mathcal{E}(f) = 1.$$

On the other hand, Examples 3.4 and 3.10 yield $R^F(-1, 1) = 2$ and $R^W(-1, 1) = 1$. Thus,

$$(R^F(x, y) - R^W(x, y))^{-1} = 1 = \mathcal{E}(h_0) = \min_{h \in H_1^{xy}} \mathcal{E}(h)$$

holds.

Chapter 5

Probabilistic Representations on Infinite Graphs

On finite graphs, the effective resistance can be written in terms of the graph's random walk in a very succinct matter, see (2.17) and (2.19). Below we will see that these representations also hold true on infinite recurrent graphs but may fail on transient graphs even if $\operatorname{dom}\mathcal{E} = \mathrm{Fin}$ holds (see Example 5.11).

For the transient case, we produce a probabilistic representation for the wired effective resistance in terms of the graph's Green's function, see Corollary 5.10, and completely characterize under which conditions the formulae from the finite case also apply to the free effective resistance, see Theorems 5.25 and 5.26.

Finally, applying martingale theory to unit potentials ϕ_D^{xy} corresponding to resistance metrics R_D of G yields

$$R_D(x, y) = \mathbf{E}_x\left[\mathbf{1}_{\{\tau_y=\infty\}}\phi_D^{xy}(X_\infty)\right] + \frac{1}{c_x}\mathbf{E}_x\left[V_x^{\tau_y-1}\right], \tag{5.1}$$

which holds on any graph, see Theorem 5.33.

The content of this chapter is in big parts due to joint work with Bachmann. Section 5.3 in particular has been reported in [41] and is currently being peer-reviewed.

Let $G = (V, c)$ be an infinite graph, $(V_n)_{n\in\mathbb{N}}$ a finite exhaustion and G_n, V_n^W and G_n^W as in sections 3.2 and 3.3. We denote by $\mathbf{P}_x$, $\mathbf{P}_x^n$ and $\mathbf{P}_x^{W,n}$ the distributions of the random walks on G, G_n and G_n^W, respectively. Their transition probabilities are denoted by p, p_n and p_n^W.

5.1 The Recurrent Case

We interpret $\mathbf{P}_x^n$ as probability measures on $\Omega = V^{\mathbb{N}_0}$ by defining $p_n(x, y) = 0$ whenever $x \notin V_n$ or $y \notin V_n$. Then,

$$p_n(x, y) = \frac{c_n(x, y)}{(c_n)_x} = \frac{c(x, y)}{c(x, V_n)}$$

for all $x, y \in V_n$. Note that

$$p_n(x, y) = p(x, y) \cdot \frac{c_x}{(c_n)_x} = p(x, y) \cdot \left(1 + \frac{c(x, V \setminus V_n)}{c(x, V_n)}\right) \geq p(x, y).$$

Proposition 5.1 (Barlow [3, Lemma 2.61]). *For $x, y \in V$, $x \neq y$, the effective resistance R^F satisfies*

$$\frac{1}{c_x \cdot \mathbf{P}_x[\tau_y \leq \tau_x^+]} \leq R^F(x, y) \leq \frac{1}{c_x \cdot \mathbf{P}_x[\tau_y < \tau_x^+]} . \tag{5.2}$$

Proof. Since $p_n(x, y) \to p(x, y)$, it follows that $\mathbf{P}_x^n$ converges weakly to $\mathbf{P}_x$. Hence, by the Portmanteau theorem [11, Theorem 3.9.1], we have

$$\limsup_{n \to \infty} \mathbf{P}_x^n(C) \leq \mathbf{P}_x(C)$$

for all closed sets $C \subseteq \Omega$ and

$$\liminf_{n \to \infty} \mathbf{P}_x^n(U) \geq \mathbf{P}_x(U)$$

for all open sets $U \subseteq \Omega$.

The sets $V \setminus \{x\}$, $V \setminus \{x, y\}$ and $\{y\}$ are open in the discrete topology on V. Hence,

$$\{\omega \in \Omega \mid \tau_y < \tau_x^+\} = \left(\{y\} \times V^{\mathbb{N}}\right) \cup \left(\bigcup_{n=1}^{\infty} (V \setminus \{y\}) \times (V \setminus \{x, y\})^{n-1} \times \{y\} \times V^{\mathbb{N}}\right)$$

is open and it follows that

$$R^F(x, y) = \lim_{n \to \infty} R_{G_n}(x, y) = \lim_{n \to \infty} \frac{1}{(c_n)_x \cdot \mathbf{P}_x^n[\tau_y < \tau_x^+]}$$

$$= \frac{1}{c_x \cdot \lim_{n \to \infty} \mathbf{P}_x^n[\tau_y < \tau_x^+]} \leq \frac{1}{c_x \cdot \mathbf{P}_x[\tau_y < \tau_x^+]} .$$

Since it is the complement of an open set, $\{\omega \in \Omega \mid \tau_y \leq \tau_x^+\}$ is closed and we have

$$R^F(x, y) = \lim_{n \to \infty} R_{G_n}(x, y) = \lim_{n \to \infty} \frac{1}{(c_n)_x \cdot \mathbf{P}_x^n[\tau_y \leq \tau_x^+]}$$

$$= \frac{1}{c_x \cdot \lim_{n \to \infty} \mathbf{P}_x^n[\tau_y \leq \tau_x^+]} \geq \frac{1}{c_x \cdot \mathbf{P}_x[\tau_y \leq \tau_x^+]} .$$

$\square$

Corollary 5.2. *If G is recurrent, then there exists only one resistance metric of G and it is given by*

$$R(x, y) = \frac{1}{c_x \cdot \mathbf{P}_x[\tau_y < \tau_x^+]} = \frac{1}{c_x \cdot \mathbf{P}_x[\tau_y \leq \tau_x^+]} = \frac{1}{c_x} \mathbf{F}_x \left[\sum_{k=0}^{\tau_y - 1} \mathbf{1}_x(X_k)\right] \tag{5.3}$$

Proof. If G is recurrent, we have $\mathbf{P}_x[\tau_x^+ = \tau_y = \infty] = 0$. Hence, (5.2) becomes

$$\frac{1}{c_x \cdot \mathbf{P}_x[\tau_y \leq \tau_x^+]} = R^F(x, y) = \frac{1}{c_x \cdot \mathbf{P}_x[\tau_y < \tau_x^+]} .$$

Since G is recurrent, it follows by Corollary 3.20 that Fin = dom $\mathcal{E}$. Hence, there exists only one resistance metric of G and it is equal to R^F. The third equality of (5.3) holds since Φ^{xy} is geometrically distributed with parameter $\mathbf{P}_x[\tau_x^+ < \tau_y]$. This can be proven exactly as in the finite case (cf. Corollary 2.19).

$\square$

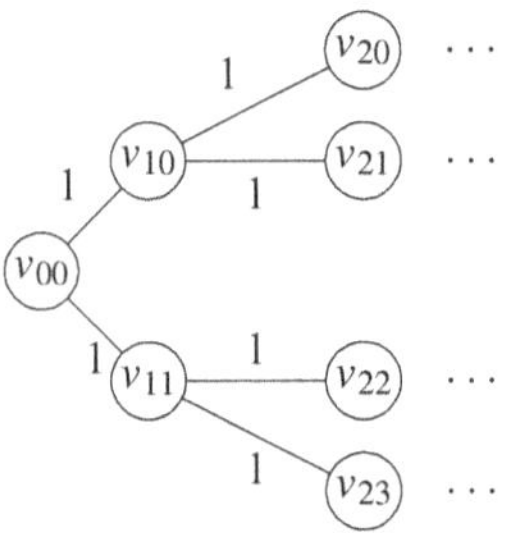

Figure 5.1: The binary tree T_2.

Another consequence of the bounds in (5.2) is an accessible sufficient condition in terms of R^F for a graph G to be transient.

Corollary 5.3. *Let G be any graph. If there exists a finite exhaustion $(V_n)_{n\in\mathbb{N}}$, $x \in V_0$, $v_n \in V \setminus V_n$ such that*

$$\exists\, K > 0\ \forall\, n \in \mathbb{N} : R^F(x, v_n) \leq K, \tag{5.4}$$

then G is transient.

Proof. We have

$$\mathbf{P}_x[\tau_{V \setminus V_n} \leq \tau_x^+] \geq \mathbf{P}_x[\tau_{v_n} \leq \tau_x^+] \geq \frac{1}{c_x \cdot R^F(x, v_n)} \geq \frac{1}{c_x \cdot K}.$$

Hence, $\lim_{n\to\infty} \mathbf{P}_x[\tau_{V \setminus V_n} \leq \tau_x^+] > 0$. Since $\mathbf{1}_{\{\tau_x^+ < \tau_{V \setminus V_n}\}} \nearrow \mathbf{1}_{\{\tau_x^+ < \infty\}}$ almost everywhere, it follows by the Monotone Convergence Theorem that

$$\mathbf{P}_x[\tau_x^+ = \infty] = 1 - \mathbf{P}_x[\tau_x^+ < \infty] = 1 - \lim_{n\to\infty} \mathbf{P}_x[\tau_x^+ < \tau_{V \setminus V_n}] = \lim_{n\to\infty} \mathbf{P}_x[\tau_{V \setminus V_n} \leq \tau_x^+] > 0.$$

Hence, G is transient. $\square$

Remark 5.4. The condition in Corollary 5.3 is satisfied if there exists $v_n \in V \setminus V_n$ such that

$$\exists\, K > 0\ \forall\, n \in \mathbb{N} : d_G(x, v_n) \leq K. \tag{5.5}$$

This is usually quite easy to check for a specific graph.

Example 5.5. We show that the condition in 5.3 is not necessary. Consider the unweighted binary tree T_2, see Figure 5.1, i.e. all existing edges have weight 1. The random walk on T_2 is transient since at any vertex the probability to jump further to the right is $2/3$. However, T_2 does not satisfy the criterion of Corollary 5.3. Indeed, let $(V_n)_{n\in\mathbb{N}}$ be $V_n := \{v_{ab} \in V \mid a \leq n\}$. Then, for $v_n \in V \setminus V_n$, we have $R^F(v_{00}, v_n) \geq R^F(v_{00}, v_{(n+1)0}) = n + 1 \to \infty$.

5.2 Wired Effective Resistance

We define

$$g_n^x(z) := \mathbf{E}_z^{W,n}[V_x^{\tau_{v_\infty}-1}] \tag{5.6}$$

to be the expected number of times the random walk on G_n^W starting in z visits x before reaching v_∞.

Lemma 5.6. *For all $x, z \in V$, we have*

$$\lim_{n\to\infty} g_n^x(z) = g(z,x) = \mathbf{E}_z[V_x].$$

Proof. The transition probabilities of $\mathbf{P}_x$ and $\mathbf{P}_x^{W,n}$ coincide in V_n and, for $v \in V_n$, we have

$$\mathbf{P}_v[X_1 \notin V_n] = \frac{c(v, V \setminus V_n)}{c_v} = \mathbf{P}_v^{W,n}[X_1 = v_\infty].$$

Hence,

$$\mathbf{E}_z[V_x^{\tau_{V \setminus V_n}}] = \mathbf{E}_z^{W,n}[V_x^{\tau_{v_\infty}-1}] = g_n^x(z) \tag{5.7}$$

and the claim follows by the Monotone Convergence Theorem [34, Theorem 11.1]. $\square$

Lemma 5.7. *For all $x \in V_n$, we have*

$$\Delta_n^W g_n^x = \mathbf{1}_x - \frac{c_x}{(c_n^W)_{v_\infty}} \cdot \mathbf{1}_{v_\infty}.$$

Proof. Since $x \in V_n$, we have $x \neq v_\infty$ and thus $V_x^{\tau_{v_\infty}-1} = V_x^{\tau_{v_\infty}}$. For $y \in V_n$, we use the Markov property (MP2) to compute

$$g_n^x(y) = \mathbf{E}_z^{W,n}\left[\sum_{k=0}^{\tau_{v_\infty}-1} \mathbf{1}_x(X_k)\right] = \mathbf{1}_x(y) + \mathbf{E}_z^{W,n}\left[\sum_{k=1}^{\tau_{v_\infty}} \mathbf{1}_x(X_k)\right]$$
$$= \mathbf{1}_x(y) + \mathbf{E}_z^{W,n}\left[\mathbf{E}_{X_1}\left[V_x^{\tau_{v_\infty}-1}\right]\right] = \mathbf{1}_x(y) + \mathbf{E}_z^{W,n}\left[g_n^x(X_1)\right].$$

Hence, $(\Delta_n^W g_n^x)(y) = \mathbf{1}_x(y)$. Since G_n^W is finite, we have $\mathbf{1} \in l^2(V_n^W)$ and thus

$$0 = \langle \nabla g_n^x, \nabla \mathbf{1}\rangle_E = \langle \Delta_n^W g_n^x, \mathbf{1}\rangle_V = \sum_{v \in V_n^W} (c_n^W)_v \cdot (\Delta_n^W g_n^x)(v) = (c_n^W)_x + (c_n^W)_{v_\infty}(\Delta_n^W g_n^x)(v_\infty).$$

The claim follows by $(c_n^W)_x = c_x$. $\square$

Corollary 5.8. *The unit potential $\phi_{W,n}^{xy}$ on G_n^W is*

$$\phi_{W,n}^{xy} = \frac{1}{c_x}g_n^x - \frac{1}{c_y}g_n^y - \left(\frac{1}{c_x}g_n^x(y) - \frac{1}{c_y}g_n^y(y)\right). \tag{5.8}$$

Proof. Let

$$\phi := \frac{1}{c_x} g_n^x - \frac{1}{c_y} g_n^y - \left(\frac{1}{c_x} g_n^x(y) - \frac{1}{c_y} g_n^y(y) \right).$$

Then,

$$\Delta\phi = \frac{1}{c_x} \mathbf{1}_x - \frac{1}{c_y} \mathbf{1}_y$$

and $\phi(y) = 0$. Hence, $\phi_{W,n}^{x,y} = \phi$ follows by the uniqueness of potentials on finite graphs (see Proposition 2.2). $\qquad\square$

Theorem 5.9.

$$\phi_{\mathrm{Fin}}^{xy}(z) = \frac{1}{c_x} g(z,x) - \frac{1}{c_y} g(z,y) - \left(\frac{1}{c_x} g(y,x) - \frac{1}{c_y} g(y,y) \right). \tag{5.9}$$

Proof. Let $\psi_n := \phi_{W,n}^{xy} / \mathcal{E}_n^W(\phi_{W,n}^{xy})$. Then, $\psi_n(x) = 1, \psi_n(y) = 0$ and $R_n^W(x,y) = \mathcal{E}_n^W(\psi_n)^{-1}$. We have

$$\psi_{\mathrm{Fin}}^{xy}(y) = 0 = \lim_{n\to\infty} \psi_n(y)$$

and, by Proposition 4.58,

$$\mathcal{E}(\psi_{\mathrm{Fin}}^{xy}) = R_{\mathrm{Fin}}(x,y)^{-1} = R^W(x,y)^{-1} = \lim_{n\to\infty} R_n^W(x,y)^{-1}$$
$$= \lim_{n\to\infty} \mathcal{E}_n^W(\psi_n) = \lim_{n\to\infty} \mathcal{E}(L_n^W(\psi_n)).$$

Fin_1^{xy} is a convex, closed set in H_y, $L_n^W(\psi_n) \in l_0(V) \subseteq \mathrm{Fin}_1^{xy}$ and

$$\lim_{n\to\infty} \| L_n^W(\psi_n) \|_{\mathrm{H}_y} = \lim_{n\to\infty} \left(\mathcal{E}(L_n^W(\psi_n)) + \psi_n(y) \right)^{\frac{1}{2}} = \| \psi_{\mathrm{Fin}}^{xy} \|_{\mathrm{H}_y}.$$

Since ψ_{Fin}^{xy} has minimal norm in Fin_1^{xy}, it follows by Theorem 1.19 that $L_n^W(\psi_n) \to \psi_{\mathrm{Fin}}^{xy}$ in H_y. In particular, we have point-wise convergence (see Prop. 1.17)

$$\phi_{\mathrm{Fin}}^{xy}(z) = \frac{\psi_{\mathrm{Fin}}^{xy}(z)}{\mathcal{E}(\psi_{\mathrm{Fin}}^{xy})} = \lim_{n\to\infty} \frac{\psi_n(z)}{\mathcal{E}_n^W(\psi_n)} = \lim_{n\to\infty} \phi_{W,n}^{xy}(z)$$
$$= \frac{1}{c_x} g(z,x) - \frac{1}{c_y} g(z,y) - \left(\frac{1}{c_x} g(y,x) - \frac{1}{c_y} g(y,y) \right)$$

by Corollary 5.8 and Lemma 5.6. $\qquad\square$

Corollary 5.10. *Let G be transient and $x, y \in V$, $x \neq y$. Then,*

$$R^W(x,y) = \frac{1}{c_x} \left(g(x,x) - g(y,x) \right) - \frac{1}{c_y} \left(g(x,y) - g(y,y) \right). \tag{5.10}$$

Proof. Follows immediately from

$$R^W(x,y) = R_{\mathrm{Fin}}(x,y) = \phi_{\mathrm{Fin}}^{xy}(x)$$

and Theorem 5.9. $\qquad\square$

Let $H_1^{xy} = \{h \in \text{Harm} \mid h(x) = 1, h(y) = 0\}$ as in Section 4.8 and R_D be a resistance metric of G. Combining corollaries 5.10 and 4.69, we get

$$R_D(x, y) = \frac{1}{c_x}\left(\mathbf{E}_x[V_x] - \mathbf{E}_y[V_x]\right) - \frac{1}{c_y}\left(\mathbf{E}_x[V_y] - \mathbf{E}_y[V_y]\right) \tag{5.11}$$

if $H_1^{xy} = \emptyset$ and

$$R_D(x, y) = \frac{1}{c_x}\left(\mathbf{E}_x[V_x] - \mathbf{E}_y[V_x]\right) - \frac{1}{c_y}\left(\mathbf{E}_x[V_y] - \mathbf{E}_y[V_y]\right) + \frac{1}{\mathcal{E}(h_D^{xy})} \tag{5.12}$$

otherwise where h_D^{xy} is an energy minimizer in $H_1^{xy} \cap D$.

5.3 Free Effective Resistance

Combining corollaries 3.13 and 3.15 in [19] yields the claim that

$$R^F(x, y) = \frac{1}{c_x \cdot \mathbf{P}_x[\tau_y < \tau_x^+]}$$

holds on any transient graph. This is false as Example 5.11 shows. However, it raises the question under which conditions this is true. More precisely, we will not investigate when the bounds in (5.2) are attained for all $x, y \in V$. Since

$$R_{G_n}(x, y) = \frac{1}{(c_n)_x \mathbf{P}_x^n[\tau_y < \tau_x^+]}$$

for all $n \in \mathbb{N}$ and $c_x = \lim_{n\to\infty}(c_n)_x$, the upper bound of (5.2) is attained if and only if

$$\lim_{n\to\infty} \mathbf{P}_x^n[\tau_y < \tau_x^+] = \mathbf{P}_x[\tau_y < \iota_x^+]. \tag{5.13}$$

Analogously, the lower bound is attained if and only if

$$\lim_{n\to\infty} \mathbf{P}_x^n[\tau_y < \tau_x^+] = \mathbf{P}_x[\tau_y \leq \tau_x^+]. \tag{5.14}$$

We will now give an example of a graph on which both inequalities in (5.2) are strict.

Example 5.11 (The transient $\mathcal{T}$). Consider the graph $\mathcal{T}$ shown in Figure 5.2. It is transient and we have $R^F(B, T) = 2$. However,

$$\begin{aligned}
\mathbf{P}_B[\tau_T < \tau_B^+] &= \mathbf{P}_0[\tau_T < \tau_B] \\
&= 1 - \mathbf{P}_0[\tau_B \leq \tau_T] \\
&= 1 - \mathbf{P}_0[\tau_B < \tau_T] - \mathbf{P}_0[\tau_B = \tau_T = \infty].
\end{aligned}$$

Due to the symmetry of $\mathcal{T}$ we have $\mathbf{P}_0[\tau_B < \tau_T] = \mathbf{P}_0[\tau_T < \tau_B]$. Together with the transience of $\mathcal{T}$, this implies

$$\mathbf{P}_B[\tau_T < \tau_B^+] = \mathbf{P}_0[\tau_T < \tau_B] = \frac{1 - \mathbf{P}_0[\tau_B = \tau_T = \infty]}{2} < \frac{1}{2}$$

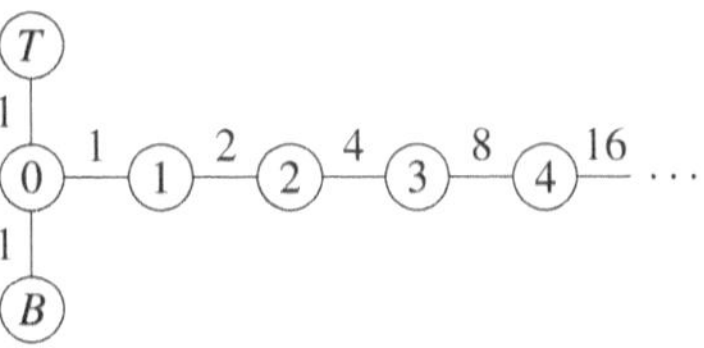

Figure 5.2: The transient graph $\mathcal{T}$

and

$$\mathbf{P}_B[\tau_T \leq \tau_B^+] = \mathbf{P}_0[\tau_T \leq \tau_B] = \frac{1 + \mathbf{P}_0[\tau_B = \tau_T = \infty]}{2} > \frac{1}{2}.$$

More precisely, one can compute

$$\mathbf{P}_B[\tau_T < \tau_B^+] = \frac{2}{5} \text{ and } \mathbf{P}_B[\tau_T \leq \tau_B^+] = \frac{3}{5}.$$

Hence,

$$\frac{1}{c_B \mathbf{P}_B[\tau_T < \tau_B^+]} \neq R^F(B, T)$$

and

$$\frac{1}{c_B} \mathbf{E}_B \left[\sum_{k=0}^{\tau_T - 1} \mathbf{1}_B(X_k) \right] = \frac{1}{c_B \mathbf{P}_B[\tau_T \leq \tau_B^+]} \neq R^F(B, T).$$

Remark 5.12. Note that, although $\mathcal{T}$ is transient, we have dom $\mathcal{E}$ = Fin since every harmonic function is constant. This shows that whether

$$R^F(x, y) = \frac{1}{c_x \cdot \mathbf{P}_x[\tau_y < \tau_x^+]}$$

holds for all $x, y \in V$ is more tightly connected to the transience of the graph than to the existence of non-trivial harmonic functions.

To check whether (5.13) and (5.14) hold, it is useful to write both sides as sums of probabilities of paths.

The probability of a path γ with respect to $\mathbf{P}_x$ is defined by

$$\mathbf{P}_x(\gamma) := \mathbf{P}_x(\{\gamma\} \times V^{\mathbb{N}}) = \mathbf{1}_x(\gamma_0) \cdot \prod_{k=0}^{L(\gamma)-1} p(\gamma_k, \gamma_{k+1}) \cdot$$

Recall that, for $A \subseteq V$, we denote by $\Gamma_G(x, y; A)$ be the set of all paths $x \to y$ in G which only use vertices in A. Using this notion and $\Gamma_{G_n}(x, y) = \Gamma_G(x, y; V_n)$, we see that (5.13) becomes

$$\lim_{n \to \infty} \sum_{\gamma \in \Gamma_G(x,y;V_n)} \mathbf{P}_x^n(\gamma) = \sum_{\gamma \in \Gamma_G(x,y)} \mathbf{P}_x(\gamma) \tag{5.15}$$

Since $\Gamma_G(x, y; V_n)$ increases to $\Gamma_G(x, y)$, this might look like an easy application of either the Monotone Convergence Theorem or the Dominated Convergence Theorem. However, both are not applicable since $\mathbf{P}_x^n(\gamma)$ may be strictly greater than $\mathbf{P}_x(\gamma)$.

To investigate when exactly (5.15) holds, we will introduce another random walk on V which can be considered an intermediary between $\mathbf{P}_x^n$ and $\mathbf{P}_x$.

The difference in the behavior of $\mathbf{P}_x$ and $\mathbf{P}_x^n$ occurs only when $\mathbf{P}_x$ leaves V_n. Instead, $\mathbf{P}_x^n$ is basically reflected back to a vertex in V_n. We will now construct an intermediary random walk which still has a finite state space, models the behavior of stepping out of V_n and has the same transition probabilities as $\mathbf{P}_x$ in V_n. This is done by adding **boundary vertices** to G_n wherever there is an edge from V_n to $V \setminus V_n$.

For any set $A \subseteq V$, let

$$\partial_i A := \{v \in A \mid \exists\, w \in V \setminus A : c(v, w) > 0\}$$

be the **inner boundary** and $\partial_o A := \partial_i(V \setminus A)$ be the **outer boundary** of A in G. For any $v \in \partial_i A$, let $\bar{v}$ be a copy of v.

Definition 5.13 (Extended random walk)**.** Define $\overline{G_n} = (\overline{V_n}, \overline{c_n})$ where

$$\overline{V_n} = V_n \,\dot{\cup}\, \{\bar{v} \mid v \in \partial_i V_n\}\,,$$

and $\overline{c_n}$ is defined as follows. For $x, y \in \overline{V_n}$, let

$$\overline{c_n}(x, y) = \overline{c_n}(y, x) = \begin{cases} c(x, y) & , \ x, y \in V_n \\ c(x, V \setminus V_n) & , \ y = \bar{x} \\ 0 & , \ \text{otherwise} \end{cases} \ .$$

In particular, we have $(\overline{c_n})_x = c_x$ for all $x \in V$. We denote by $\overline{\mathbf{P}_x^n}$ the random walk on $\overline{G_n}$ starting in x with transition matrix $\overline{p_n}$ given by

$$\overline{p_n}(x, y) = \frac{\overline{c_n}(x, y)}{(c_n)_x}\,,$$

Furthermore, define $V_n^* := \overline{V_n} \setminus V_n$.

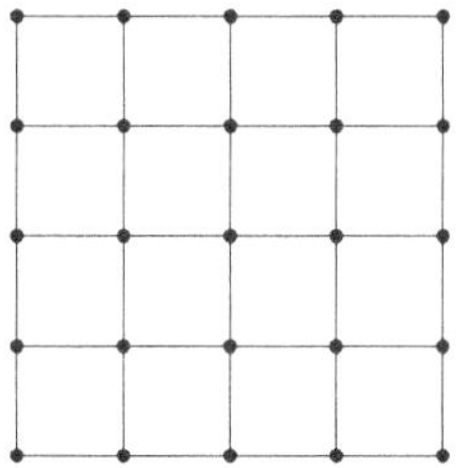

Figure 5.3: The lattice $\mathbb{Z}^2$.

Example 5.14. Let G be the lattice $\mathbb{Z}^2$ with unit weights, see Figure 5.3. Furthermore, let $V_n := \{-n\ldots,0,\ldots,n\}^2$. G_1 and $\overline{G_1}$ are illustrated in Figure 5.4. Note that $c((1,1),\overline{(1,1)}) = 2$ since $(1,1)$ has two edges leaving V_1 in G.

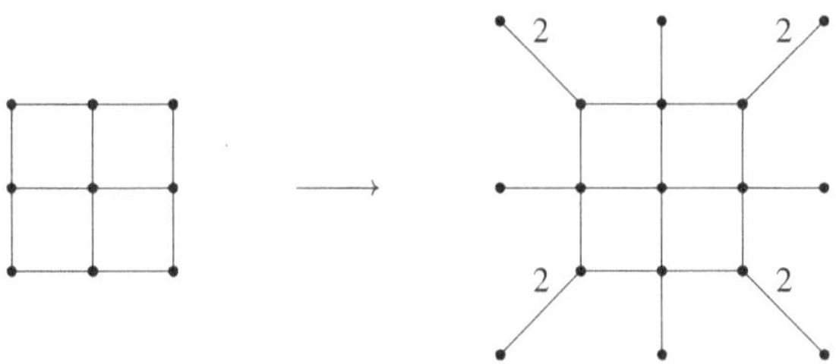

Figure 5.4: G_1 (left) and $\overline{G_1}$ (right) for $G = \mathbb{Z}^2$.

Lemma 5.15 (Relation of $p_n, \overline{p_n}$ and p). *For $x, y \in V_n$ we have*

$$p_n(x,y) \geq p(x,y) = \overline{p_n}(x,y).$$

For $x, y \in V$ and $m \in \mathbb{N}$ such that $x, y \in V_m$, we have

$$\lim_{n\to\infty} p_n(x,y) = p(x,y) = \overline{p_m}(x,y).$$

Note that for $n \in \mathbb{N}$ and $x, y \in V_n$, we have

$$\Gamma_{G_n}(x,y) = \Gamma_{\overline{G_n}}(x,y;V_n) = \Gamma_G(x,y;V_n).$$

By Lemma 5.15, the following holds for all $x, y \in V_n$.

$$\forall\, m \geq n\ \forall\, \gamma \in \Gamma_G(x,y;V_m) : \mathbf{P}_x(\gamma) = \overline{\mathbf{P}_x^m}(\gamma).$$

The connection between $\mathbf{P}_x^n(\gamma)$ and $\overline{\mathbf{P}_x^n}(\gamma)$ is a bit more intricate. In order to investigate this connection, first consider what kind of paths exist in $\overline{G_n}$. Let $x, y \in V_n$ and $\overline{\gamma} \in \Gamma_{\overline{G_n}}(x,y)$ with $l = L(\gamma) \geq 2$. Then, by Definition of $\overline{G_n}$, there exist $v_1,\ldots,v_{l-1} \in V_n$, $k_1,\ldots,k_{l-1} \in \mathbb{N}_0$ such that

$$\overline{\gamma} = (x, (v_1)_{k_1},\ldots,(v_{n-1})_{k_{l-1}}, y) \tag{5.16}$$

where $(v)_k := (v,\overline{v},v,\ldots,\overline{v},v)$ for $v \in \partial_i V_n$ and $k \in \mathbb{N}_0$ or $v \in V_n \setminus \partial_i V_n$ and $k = 0$. Note
$$\underbrace{\phantom{(v,\overline{v},v,\ldots,\overline{v},v)}}_{k\ \text{times}}$$
that the representation (5.16) is unique for $\overline{\gamma}$ since G_n does not contain any self-loops.

Definition 5.16. For $x, y \in V_n$, let $\pi : \Gamma_{\overline{G_n}}(x,y) \to \Gamma_{G_n}(x,y)$ be the projection of $\Gamma_{\overline{G_n}}(x,y)$ onto $\Gamma_{G_n}(x,y)$ which replaces all occurrences of $(v,\overline{v},v)$ for any $v \in \partial_i V_n$ by (v).

More precisely, let $\overline{\gamma} \in \Gamma_{\overline{G_n}}(x,y)$. If $L(\overline{\gamma}) = 1$, then $\overline{\gamma} = (x,y)$ and we define $\pi(\overline{\gamma}) := (x,y)$. If $L(\overline{\gamma}) \geq 2$, it is of the form (5.16) and we define

$$\pi(\overline{\gamma}) := (x, v_1,\ldots,v_{l-1}, y) \tag{5.17}$$

Lemma 5.17. *For all $x \neq y$ and $\gamma \in \Gamma_{G_n}(x, y)$, we have*

$$\mathbf{P}_x^n(\gamma) = \frac{c_x}{(c_n)_x} \cdot \sum_{\overline{\gamma} \in \pi^{-1}(\gamma)} \overline{\mathbf{P}_x^n}(\overline{\gamma}).$$

Proof. For any $v, w \in V_n$, we have

$$\frac{c_v}{(c_n)_v} \cdot \overline{p_n}(v, w) = \frac{c_v}{(c_n)_v} \cdot \frac{c(v, w)}{c_v} = p_n(v, w) \tag{5.18}$$

and, if $v \in \partial_i V_n$,

$$\sum_{k=0}^{\infty} (\underbrace{\overline{p_n}(v, \overline{v}) \cdot \overline{p_n}(\overline{v}, v)}_{=1})^k = \frac{1}{1 - \overline{p_n}(v, \overline{v})} = \frac{c_v}{c_v - \sum_{w \notin V_n} c(v, w)} = \frac{c_v}{(c_n)_v} \cdot \tag{5.19}$$

For $\gamma \in \Gamma_{G_n}(x, y)$ with $L(\gamma) = 1$, we have $\gamma = (x, y)$. Since any $\overline{\gamma} \in \pi^{-1}(\gamma)$ visits x and y only once, $\pi^{-1}(\gamma) = \{\gamma\}$ holds and

$$\frac{c_x}{(c_n)_x} \cdot \sum_{\overline{\gamma} \in \pi^{-1}(\gamma)} \overline{\mathbf{P}_x^n}(\overline{\gamma}) = \frac{c_x}{(c_n)_x} \cdot \overline{\mathbf{P}_x^n}(\gamma) = \frac{c_x}{(c_n)_x} \cdot \overline{p_n}(x, y) = p_n(x, y) = \mathbf{P}_x^n(\gamma).$$

Now let $\gamma = (x, v_1, \ldots, v_{l-1}, y) \in \Gamma_{G_n}(x, y)$ with $l = L(\gamma) \geq 2$. We define

$$A(\gamma) := \left\{ (k_1, \ldots, k_{l-1}) \in (\mathbb{N}_0)^{l-1} \mid \text{for each } j \in \{1, \ldots, l-1\} : k_j = 0 \text{ if } v_j \notin \partial_i V_n \right\}.$$

It follows that

$$\pi^{-1}(\gamma) = \left\{ \overline{\gamma} \in \Gamma_{\overline{G_n}}(x, y) \mid \pi(\overline{\gamma}) = \gamma \right\}$$
$$= \left\{ (x, (v_1)_{k_1}, \ldots, (v_{l-1})_{k_{l-1}}, y) \mid (k_1, \ldots, k_{l-1}) \in A(\gamma) \right\}$$

and we compute

$$\sum_{\overline{\gamma} \in \pi^{-1}(\gamma)} \overline{\mathbf{P}_x^n}(\overline{\gamma}) = \sum_{(k_1, \ldots, k_{l-1}) \in A(\gamma)} \overline{\mathbf{P}_x^n}((x, (v_1)_{k_1}, \ldots, (v_{l-1})_{k_{l-1}}, y))$$

$$= \sum_{(k_1, \ldots, k_{l-1}) \in A(\gamma)} \left[\overline{\mathbf{P}_x^n}((x, v_1, \ldots, v_{l-1}, y)) \cdot \prod_{\substack{j=1, \ldots, l-1 \\ v_j \in \partial_i V_n}} \overline{p_n}(v_j, \overline{v_j})^{k_j} \right]$$

$$\overset{(5.19)}{=} \overline{\mathbf{P}_x^n}((x, v_1, \ldots, v_{l-1}, y)) \cdot \prod_{j=1}^{l-1} \frac{c_{v_j}}{(c_n)_{v_j}}$$

$$\overset{(5.18)}{=} \frac{(c_n)_x}{c_x} \cdot \mathbf{P}_x^n((x, v_1, \ldots, v_{l-1}, y)) = \frac{(c_n)_x}{c_x} \cdot \mathbf{P}_x^n(\gamma)$$

$\square$

Remark 5.18. The original version of Definition 5.16 and the proof of Lemma 5.17 as stated in [41] were erroneous. The revised versions are heavily based on the referee report [2].

Proposition 5.19. *For $x, y \in V_n$, $x \neq y$, we have*

$$\overline{\mathbf{P}_x^n}[\tau_y < \tau_x^+] = \frac{(c_n)_x}{c_x} \cdot \mathbf{P}_x^n[\tau_y < \tau_x^+].$$

Proof. Using

$$\Gamma_{\overline{G_n}}(x, y) = \bigcup_{\gamma \in \Gamma_{G_n}(x,y)}^{\cdot} \pi^{-1}(\gamma),$$

we compute

$$\begin{aligned}
\mathbf{P}_x^n[\tau_y < \tau_x^+] &= \sum_{\gamma \in \Gamma_{G_n}(x,y)} \mathbf{P}_x^n(\gamma) = \sum_{\gamma \in \Gamma_{G_n}(x,y)} \left(\frac{c_x}{(c_n)_x} \cdot \sum_{\overline{\gamma} \in \pi^{-1}(\gamma)} \overline{\mathbf{P}_x^n}(\overline{\gamma}) \right) \\
&= \frac{c_x}{(c_n)_x} \cdot \sum_{\overline{\gamma} \in \Gamma_{\overline{G_n}}(x,y)} \overline{\mathbf{P}_x^n}(\overline{\gamma}) = \frac{c_x}{(c_n)_x} \cdot \overline{\mathbf{P}_x^n}[\tau_y < \tau_x^+].
\end{aligned}$$

$\square$

Since we now have clarified the relation between $\mathbf{P}_x^n$, $\overline{\mathbf{P}_x^n}$ and $\mathbf{P}_x$, we can return our attention to (5.13).

Proposition 5.20. *For $x, y \in V$, $x \neq y$, we have*

$$\lim_{n \to \infty} \mathbf{P}_x^n[\tau_y < \tau_x^+] = \mathbf{P}_x[\tau_y < \tau_x^+]$$

if and only if

$$\lim_{n \to \infty} \overline{\mathbf{P}_x^n}[\tau_{V_n^*} < \tau_y < \tau_x^+] = 0. \tag{5.20}$$

Proof. We have

$$\begin{aligned}
\mathbf{P}_x[\tau_y < \tau_x^+] &= \sum_{\gamma \in \Gamma_G(x,y)} \mathbf{P}_x(\gamma) = \lim_{n \to \infty} \sum_{\gamma \in \Gamma_G(x,y;V_n)} \mathbf{P}_x(\gamma) \tag{5.21} \\
&= \lim_{n \to \infty} \sum_{\gamma \in \Gamma_G(x,y;V_n)} \overline{\mathbf{P}_x^n}(\gamma)
\end{aligned}$$

and

$$\mathbf{P}_x^n[\tau_y < \tau_x^+] = \frac{c_x}{(c_n)_x} \cdot \overline{\mathbf{P}_x^n}[\tau_y < \tau_x^+] = \frac{c_x}{(c_n)_x} \cdot \sum_{\gamma \in \Gamma_{\overline{G_n}}(x,y)} \overline{\mathbf{P}_x^n}(\gamma). \tag{5.22}$$

Since $\Gamma_G(x, y; V_n) = \Gamma_{\overline{G_n}}(x, y; V_n)$ and $(c_n)_x \to c_x$, it follows that $\mathbf{P}_x^n[\tau_y < \tau_x^+] \to \mathbf{P}_x[\tau_y < \tau_x^+]$ holds if and only if

$$\lim_{n \to \infty} \sum_{\substack{\gamma \in \Gamma_{\overline{G_n}}(x,y) \\ \gamma \notin \Gamma_{\overline{G_n}}(x,y;V_n)}} \overline{\mathbf{P}_x^n}(\gamma) = 0.$$

This is the same as

$$\lim_{n \to \infty} \overline{\mathbf{P}_x^n}[\tau_{V_n^*} < \tau_y < \tau_x^+] = 0.$$

$\square$

Using the same approach, we can also characterize when (5.14) holds.

Proposition 5.21. *For $x, y \in V$, $x \neq y$, we have*

$$\lim_{n \to \infty} \mathbf{P}_x^n[\tau_y < \tau_x^+] = \mathbf{P}_x[\tau_y \leq \tau_x^+]$$

if and only if

$$\lim_{n \to \infty} \overline{\mathbf{P}_x^n}[\tau_{V_n^*} < \tau_y < \tau_x^+] = \mathbf{P}_x[\tau_x^+ = \tau_y = \infty] \tag{5.23}$$

which in turn is equivalent to

$$\lim_{n \to \infty} \overline{\mathbf{P}_x^n}[\tau_{V_n^*} < \tau_x^+ < \tau_y] = 0. \tag{5.24}$$

Proof. Using (5.21) and (5.22) from the proof of Proposition 5.20, we have

$$\mathbf{P}_x[\tau_y \leq \tau_x^+] = \mathbf{P}_x[\tau_x^+ = \tau_y = \infty] + \lim_{n \to \infty} \sum_{\gamma \in \Gamma_{\overline{G_n}}(x,y;V_n)} \overline{\mathbf{P}_x^n}(\gamma)$$

and

$$\lim_{n \to \infty} \mathbf{P}_x^n[\tau_y < \tau_x^+] = \lim_{n \to \infty} \sum_{\gamma \in \Gamma_{\overline{G_n}}(x,y)} \overline{\mathbf{P}_x^n}(\gamma)$$

provided either one of these two limits exists.

Hence, we have convergence as desired if and only if

$$\mathbf{P}_x[\tau_x^+ = \tau_y = \infty] = \lim_{n \to \infty} \sum_{\substack{\gamma \in \Gamma_{\overline{G_n}}(x,y) \\ \gamma \notin \Gamma_{\overline{G_n}}(x,y;V_n)}} \overline{\mathbf{P}_x^n}(\gamma) = \lim_{n \to \infty} \overline{\mathbf{P}_x^n}[\tau_{V_n^*} < \tau_y < \tau_x^+].$$

On the other hand, we have

$$\begin{aligned}
\mathbf{P}_x[\tau_x^+ = \tau_y = \infty] &= \lim_{n \to \infty} \mathbf{P}_x[\tau_{V \setminus V_n} < \min(\tau_x^+, \tau_y)] \\
&= \lim_{n \to \infty} \overline{\mathbf{P}_x^n}[\tau_{V_n^*} < \min(\tau_x^+, \tau_y)] \\
&= \lim_{n \to \infty} \left(\overline{\mathbf{P}_x^n}[\tau_{V_n^*} < \tau_x^+ < \tau_y] + \overline{\mathbf{P}_x^n}[\tau_{V_n^*} < \tau_y < \tau_x^+] \right)
\end{aligned}$$

which implies the second claim. $\square$

Remark 5.22. An equivalent approach would be to consider a *lazy* random walk on G_n which has the same transition probabilities $p(v, w)$ as $\mathbf{P}_x$ for $v \neq w$ but stays at v with probability

$$\sum_{w \in V \setminus V_n} p(v, w) = \mathbf{P}_v[\omega_1 \in V \setminus V_n].$$

In that case the notion of "stepping out of V_n" would be modeled by staying at any vertex $v \in V_n$.

We will show that whenever a graph G is transient and not part of an infinite line, one can find a subgraph of G which is similar to $\mathcal{T}$ from Section 5.11. We will also show that this is sufficient for (5.20) to be false.

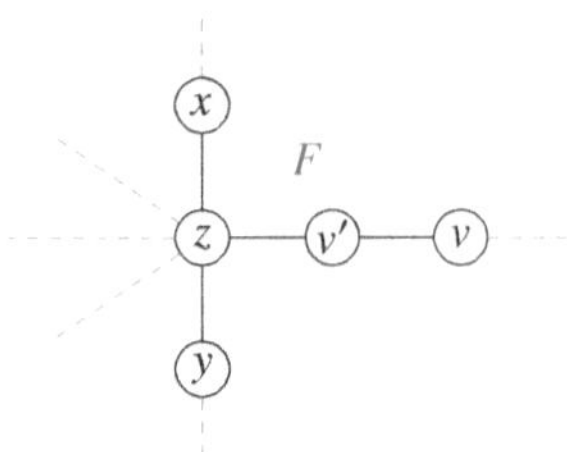

Figure 5.5: A possible choice of the set F in the proof of Proposition 5.23.

Proposition 5.23. *Let G be a transient, connected graph which is not a subgraph of a line. Then, there exist $x, y, z \in V$ such that $x \neq y$, (x, z, y) is a path in G and*

$$\mathbf{P}_z[\tau_x = \tau_y = \infty] > 0.$$

Proof. Since G is transient, it is infinite. If G is not a subgraph of a line, then there exists some $z \in V$ with at least three adjacent vertices. Let F be a set of exactly three neighbors of z. Since G is transient and F is finite, there exists $v \in \partial_o F$ such that

$$\mathbf{P}_v[\tau_F = \infty] > 0.$$

If $v = z$, we can choose $x, y \in F$, $x \neq y$, and get

$$\mathbf{P}_z[\tau_x = \tau_y = \infty] \geq \mathbf{P}_v[\tau_F = \infty] > 0.$$

If $v \neq z$, then there exists $v' \in F$ such that (z, v', v) is a path in G. Let $x, y \in V$ be such that $F = \{x, y, v'\}$, see Figure 5.5. It follows that

$$\begin{aligned}
\mathbf{P}_z[\tau_x = \tau_y = \infty] &\geq \mathbf{P}_z[X_1 = v', X_2 = v, \tau_x = \tau_y = \infty] \\
&= p(z, v') \cdot p(v', v) \cdot \mathbf{P}_v[\tau_x = \tau_y = \infty] \\
&\geq p(z, v') \cdot p(v', v) \cdot \mathbf{P}_v[\tau_F = \infty] > 0.
\end{aligned}$$

$\square$

Proposition 5.24. *Let G be a transient, connected graph. Then,*

$$\forall\, x, y \in V, \ x \neq y : \lim_{n \to \infty} \overline{\mathbf{P}_x^n}[\tau_{V_n^*} < \tau_y < \tau_x^+] = 0$$

holds if and only if G is a subgraph of an infinite line.

Proof. First, assume that G is a subgraph of an infinite line and let $x, y \in V, x \neq y$. Then, for any $n \in \mathbb{N}$ sufficiently big, we have

$$\Gamma_{\overline{G_n}}(x, y) \setminus \Gamma_{\overline{G_n}}(x, y; V_n) = \emptyset,$$

i.e. there exists no path $x \to y$ which leaves V_n before reaching y. Hence,

$$\lim_{n \to \infty} \overline{\mathbf{P}_x^n}[\tau_{V_n^*} < \tau_y < \tau_x^+] = 0.$$

To prove the converse direction, suppose that G is not a subgraph of a line. By Proposition 5.23, we know that there exist distinct vertices $x, y, z \in V$ such that (x, z, y) is a path in G and $\mathbf{P}_z[\tau_x = \tau_y = \infty] > 0$. Hence,

$$
\begin{aligned}
0 &< \mathbf{P}_z[\tau_x = \tau_y = \infty] \\
&= \lim_{n \to \infty} \mathbf{P}_z[\tau_{\partial_o V_n} < \min(\tau_x, \tau_y)] \\
&= \lim_{n \to \infty} \overline{\mathbf{P}_z^n}[\tau_{V_n^*} < \min(\tau_x, \tau_y)] \\
&= \lim_{n \to \infty} \left(\overline{\mathbf{P}_z^n}[\tau_{V_n^*} < \tau_x < \tau_y] + \overline{\mathbf{P}_z^n}[\tau_{V_n^*} < \tau_y < \tau_x] \right) \\
&\leq \limsup_n \overline{\mathbf{P}_z^n}[\tau_{V_n^*} < \tau_x < \tau_y] + \limsup_n \overline{\mathbf{P}_z^n}[\tau_{V_n^*} < \tau_y < \tau_x].
\end{aligned}
$$

Without loss of generality assume that $\limsup_{n \to \infty} \overline{\mathbf{P}_z^n}[\tau_{V_n^*} < \tau_y < \tau_x] > 0$. It follows that $\limsup_{n \to \infty} \overline{\mathbf{P}_x^n}[\tau_{V_n^*} < \tau_y < \tau_x^+] > 0$ because for all $n \in \mathbb{N}$ with $\{x, y, z\} \subseteq V_n$, we have

$$
\begin{aligned}
\overline{\mathbf{P}_x^n}[\tau_{V_n^*} < \tau_y < \tau_x^+] &\geq \overline{\mathbf{P}_x^n}[\tau_z < \tau_{V_n^*} < \tau_y < \tau_x^+] \\
&= \overline{\mathbf{P}_x^n}[\tau_z < \min(\tau_{V_n^*}, \tau_x^+, \tau_y)] \cdot \overline{\mathbf{P}_z^n}[\tau_{V_n^*} < \tau_y < \tau_x^+] \\
&\geq p(x, z) \cdot \overline{\mathbf{P}_z^n}[\tau_{V_n^*} < \tau_y < \tau_x].
\end{aligned}
$$

$\square$

Theorem 5.25. *Let G be a transient, connected graph. Then,*

$$
R^F(x, y) = \frac{1}{c_x \cdot \mathbf{P}_x[\tau_y < \tau_x^+]} \tag{5.25}
$$

holds for all $x, y \in V$ with $x \neq y$ if and only if G is a subgraph of an infinite line.

Proof. As seen in (5.13), the desired probabilistic representation (5.25) holds if and only if

$$
\lim_{n \to \infty} \mathbf{P}_x^n[\tau_y < \tau_x^+] = \mathbf{P}_x[\tau_y < \tau_x^+].
$$

By Proposition 5.20, this is equivalent to

$$
\lim_{n \to \infty} \overline{\mathbf{P}_x^n}[\tau_{V_n^*} < \tau_y < \tau_x^+] = 0
$$

and the claim follows by Proposition 5.24. $\square$

Theorem 5.26. *Let G be an infinite, connected graph. If*

$$
\forall\, x, y \in V, \; x \neq y : \lim_{n \to \infty} \overline{\mathbf{P}_x^n}[\tau_{V_n^*} < \tau_y < \tau_x^+] = \mathbf{P}_x[\tau_x^+ = \tau_y = \infty]
$$

holds, then G is recurrent.

Proof. By Proposition 5.21, we have

$$
\forall\, x, y \in V : \lim_{n \to \infty} \mathbf{P}_x[\tau_{V_n^*} < \tau_x^+ < \tau_y] = 0. \tag{5.26}
$$

Suppose that G is transient and not a subgraph of a line. Using the same arguments as in the proof of Proposition 5.24, we see that there exist distinct vertices $x, y, z \in V$ such that $(x, z, y) \in \Gamma_G(x, y)$ and

$$\limsup_n \overline{\mathbf{P}_z^n}[\tau_{V_n^*} < \tau_x < \tau_y] + \limsup_n \overline{\mathbf{P}_z^n}[\tau_{V_n^*} < \tau_y < \tau_x] > 0.$$

Since

$$\overline{\mathbf{P}_x^n}[\tau_{V_n^*} < \tau_x^+ < \tau_y] \geq p(x, z) \cdot \overline{\mathbf{P}_z^n}[\tau_{V_n^*} < \tau_x < \tau_y],$$

for all $n \in \mathbb{N}$ with $\{x, y, z\} \subseteq V_n$, it follows from (5.26) that

$$\limsup_n \overline{\mathbf{P}_z^n}[\tau_{V_n^*} < \tau_x < \tau_y] = 0$$

which implies

$$\limsup_{n \to \infty} \overline{\mathbf{P}_z^n}[\tau_{V_n^*} < \tau_y < \tau_x] > 0.$$

However, we also have

$$\overline{\mathbf{P}_y^n}[\tau_{V_n^*} < \tau_y^+ < \tau_x] \geq p(y, z) \cdot \overline{\mathbf{P}_z^n}[\tau_{V_n^*} < \tau_y < \tau_x]$$

for all $n \in \mathbb{N}$ with $\{x, y, z\} \subseteq V_n$, and it follows that

$$\limsup_n \overline{\mathbf{P}_y^n}[\tau_{V_n^*} < \tau_y^+ < \tau_x] > 0$$

which is a contradiction to (5.26).

Hence, if G is transient, then it must be a subgraph of a line. In this case,

$$\lim_{n \to \infty} \overline{\mathbf{P}_x^n}[\tau_{V_n^*} < \tau_y < \tau_x^+] = 0$$

follows for all $x, y \in V$ with $x \neq y$ by Proposition 5.24. Together with (5.26), this implies

$$\mathbf{P}_x[\tau_x^+ = \tau_y = \infty] = 0$$

for all $x, y \in V$ with $x \neq y$. However, this is a contradiction to the transience of G.

$\square$

Corollary 5.27. *Let G be an infinite, connected graph. Then,*

$$R^F(x, y) = \frac{1}{c_x \cdot \mathbf{P}_x[\tau_y \leq \tau_x^+]} \tag{5.27}$$

holds for all $x, y \in V$ with $x \neq y$ if and only if G is recurrent.

Proof. If G is recurrent, we have $\mathbf{P}_x[\tau_x^+ = \tau_y = \infty] = 0$ for all $x, y \in V$. Hence,

$$\mathbf{P}_x[\tau_x^+ < \tau_y] = \mathbf{P}_x[\tau_x^+ \leq \tau_y]$$

and (5.2) implies the claim.

If (5.27) holds for all $x, y \in V$ with $x \neq y$, then we have

$$\lim_{n \to \infty} \mathbf{P}_x^n[\tau_y < \tau_x^+] = \mathbf{P}_x[\tau_y \leq \tau_x^+]$$

for all $x, y \in V$ with $x \neq y$, and Proposition 5.21 and Theorem 5.26 imply the recurrence of G.

$\square$

This shows that the lower bound in (5.2) is actually a strict inequality for some $x, y \in V$ with $x \neq y$ for any transient graph $G = (V, c)$. In particular,

$$R^F(x, y) = \frac{1}{c_x} \mathbf{E}_x \left[\sum_{k=0}^{\tau_y - 1} \mathbf{1}_x(X_k) \right] \tag{5.28}$$

does **not** hold for all $x, y \in V$ on any transient graph.

5.4　Resistance Metrics of Graphs

In order to relate a resistance metrics R_D to the random walk of G, we will use martingale theory to investigate the limiting behavior of $\phi_D^{xy}(X_n)$. Considering $\lim_n f(X_n)$ for $f \in \operatorname{dom} \mathcal{E}$ has been done before (cf. [30, Theorem 9.11]) but we have not found any results relating to resistance metrics in the literature.

Lemma 5.28. *For any $f : V \to \mathbb{R}$, we have*

$$f(x) = \mathbf{E}_x[f(X_n)] + \sum_{k=0}^{n-1} \mathbf{E}_x[(\Delta f)(X_k)]. \tag{5.29}$$

Proof. Let $k \in \mathbb{N}$. Then,

$$\mathbf{E}_x[f(X_k) - f(X_{k-1})] = \sum_{(y,z) \in V^2} \mathbf{P}_x[X_{k-1} = y, X_k = z] \cdot (f(z) - f(y))$$

$$= \sum_{y \in V} \left[\mathbf{P}_x[X_{k-1} = y] \sum_{z \in V} \mathbf{P}_y[X_1 = z] \cdot (f(z) - f(y)) \right]$$

$$= \sum_{y \in V} \left[\mathbf{P}_x[X_{k-1} = y] \sum_{z \in V} \frac{c(y, z)}{c_y} \cdot (f(z) - f(y)) \right]$$

$$= \sum_{y \in V} \mathbf{P}_x[X_{k-1} = y] \cdot (-\Delta f)(y) = \mathbf{E}_x[(-\Delta f)(X_{k-1})].$$

Hence,

$$\mathbf{E}_x[f(X_n)] = \mathbf{E}_x[f(X_0)] + \sum_{k=1}^{n} \mathbf{E}_x[f(X_k) - f(X_{k-1})]$$

$$= f(x) - \sum_{k=1}^{n} \mathbf{E}_x[(\Delta f)(X_{k-1})]$$

and the claim follows. $\qquad\square$

Fix a resistance form $(\mathcal{E}, D)$ with $\operatorname{Fin} \subseteq D \subseteq \operatorname{dom} \mathcal{E}$ and $x, y \in V$, $x \neq y$. Define $M_0 := \phi_D^{xy}(x)$ and for $n \geq 1$,

$$M_n := \phi_D^{xy}(X_n) + \sum_{k=0}^{n-1} (\Delta \phi_D^{xy})(X_k). \tag{5.30}$$

Then, $(M_n)_{n\in\mathbb{N}}$ is a stochastic process in discrete time and it satisfies

$$M_n = \phi_D^{xy}(X_n) + \frac{1}{c_x}V_x^{n-1} - \frac{1}{c_y}V_y^{n-1} \tag{5.31}$$

by Corollary 4.64.

Proposition 5.29. *M_n is a martingale with respect to $\mathcal{F}_n := \sigma(X_0, \ldots, X_n)$.*

Proof. First, note that M_n is $\mathcal{F}_n$-measurable. By the Markov property of $\mathcal{E}$, we have $\phi_D^{xy}(V) \subseteq [0, R_D(x, y)]$. Since G is transient, we have $\mathbf{E}_x[V_x] < \infty$ and $\mathbf{E}_x[V_y] < \infty$, see Lemma 1.24. Hence,

$$\mathbf{E}_x[|M_n|] \le \mathbf{E}_x[\phi_D^{xy}(X_n)] + \frac{1}{c_x}\mathbf{E}_x[V_x^{n-1}] + \frac{1}{c_y}\mathbf{E}_x[V_y^{n-1}]$$

$$\le R_D(x, y) + \frac{1}{c_x}\mathbf{E}_x[V_x] + \frac{1}{c_y}\mathbf{E}_x[V_y] < \infty$$

for all $n \in \mathbb{N}$. We can compute

$$\mathbf{E}_x[M_{n+1} - M_n \mid \mathcal{F}_n] = \mathbf{E}_x\left[\phi_D^{xy}(X_{n+1}) \mid \mathcal{F}_n\right] + (\Delta\phi_D^{xy})(X_n) - \phi_D^{xy}(X_n)$$

$$= \mathbf{E}_x\left[\phi_D^{xy}(X_{n+1}) \mid \mathcal{F}_n\right] - \mathbf{E}_{X_n}[\phi_D^{xy}(X_1)] = 0$$

which is equivalent to $\mathbf{E}_x[M_{n+1} \mid \mathcal{F}_n] = M_n$. $\qquad\square$

Proposition 5.30. *There exists a random variable M_∞ such that $M_n \to M_\infty$ almost surely and in L^1 with respect to $\mathbf{P}_x$.*

Proof. By the Martingale Convergence Theorem (see e.g. [11, Theorem 5.5.6]), we have to show that M_n is uniformly integrable. In order to show that a family $\mathcal{X}$ of random variables is uniformly integrable, it suffices to show that $\sup_{X\in\mathcal{X}} \mathbf{E}[X^2] < \infty$.

By Lemma 1.25, we have

$$\sup_{n\in\mathbb{N}} \mathbf{E}_x\left[\left(\frac{1}{c_x}V_x^{n-1}\right)^2\right] \le \frac{1}{c_x^2}\sup_{n\in\mathbb{N}} \mathbf{E}_x[(V_x)^2] < \infty$$

and

$$\sup_{n\in\mathbb{N}} \mathbf{E}_x\left[\left(\frac{1}{c_y}V_y^{n-1}\right)^2\right] \le \frac{1}{c_y^2}\sup_{n\in\mathbb{N}} \mathbf{E}_x[(V_y)^2] < \infty.$$

Since $\phi_D^{xy}(X_n)^2 \le R_D(x, y)^2$, it follows that the families $\left\{\phi_D^{xy}(X_n)\right\}_{n\in\mathbb{N}}$, $\left\{c_x^{-1}V_x^{n-1}\right\}_{n\in\mathbb{N}}$ and $\left\{c_y^{-1}V_y^{n-1}\right\}_{n\in\mathbb{N}}$ are uniformly integrable. Hence,

$$M_n = \phi_D^{xy}(X_n) + \frac{1}{c_x}V_x^{n-1} - \frac{1}{c_y}V_y^{n-1}$$

is also uniformly integrable. $\qquad\square$

Corollary 5.31. $\phi_D^{xy}(X_\infty) := \lim_{n\to\infty} \phi_D^{xy}(X_n)$ *exists almost surely.*

Proof. Clearly, $V_x^{n-1} \to V_x$ and $V_y^{n-1} \to V_y$ almost surely. Hence,

$$\lim_{n\to\infty} \phi_D^{xy}(X_n) = \lim_{n\to\infty} M_n - \frac{1}{c_x} V_x^{n-1} + \frac{1}{c_y} V_y^{n-1}$$

exists almost surely. $\qquad\square$

Remark 5.32. While this may be the case, our notation $\phi_D^{xy}(X_\infty)$ is not meant to imply that there exists a random variable X_∞ such that $X_n \to X_\infty$.

Theorem 5.33. *Let G be a transient graph and R_D a resistance metric of G with corresponding resistance form $(\mathcal{E}, D)$. Then,*

$$R_D(x, y) = \mathbf{E}_x\left[\phi_D^{xy}(X_\infty)\right] + \frac{1}{c_x}\mathbf{E}_x[V_x] - \frac{1}{c_y}\mathbf{E}_x[V_y] \tag{5.32}$$

and

$$R_D(x, y) = \mathbf{E}_x\left[\mathbf{1}_{\{\tau_y=\infty\}}\phi_D^{xy}(X_\infty)\right] + \frac{1}{c_x}\mathbf{E}_x\left[V_x^{\tau_y-1}\right]. \tag{5.33}$$

Proof. We have

$$R_D(x, y) = \phi_D^{xy}(x) = \mathbf{E}_x[M_\infty] = \mathbf{E}_x\left[\phi_D^{xy}(X_\infty)\right] + \frac{1}{c_x}\mathbf{E}_x[V_x] - \frac{1}{c_y}\mathbf{E}_x[V_y].$$

The second equality follows analogously when considering the stopped Martingale $M_{n\wedge\tau_y}$. $\qquad\square$

Corollary 5.34.
$$\mathbf{E}_x\left[\phi_{\mathrm{Fin}}^{xy}(X_\infty)\right] = \frac{1}{c_y}\mathbf{E}_y[V_y] - \frac{1}{c_x}\mathbf{E}_y[V_x]. \tag{5.34}$$

Proof. Follows by comparing (5.32) for $D = \mathrm{Fin}$ with (5.10). $\qquad\square$

If G is recurrent, we have $\mathbf{P}_x[\tau_y = \infty] = 0$ and $R_D = R^F = R^W$. In that case, (5.33) becomes (5.3). Hence, (5.33) actually holds for any graph G, no matter whether it is transient or recurrent. This means that we can interpret $\mathbf{E}_x\left[\mathbf{1}_{\{\tau_y=\infty\}}\phi_D^{xy}(X_\infty)\right]$ as the "defect" of resistance metrics of transient graphs.

Remark 5.35. Based on the form of the "defect term" in (5.33), we suspect that the family of resistance metrics can be parameterized with some kind of boundary argument. A boundary representation for all Dirichlet forms associated to a graph for example can be achieved by using the Royden boundary [22]. In fact, resistance forms restricted to certain L^2 spaces are such Dirichlet forms [29].